A STATLIB Primer

A STATLIB Primer

The Forecasting Process Through Statistical Computing

H. Levenbach

W. M. Brelsford

J. P. Cleary

LIFETIME LEARNING PUBLICATIONS
Belmont, California
A division of Wadsworth, Inc.

London, Singapore, Sydney, Toronto, Mexico City

Printed in the United States of America

1 2 3 4 5 6 7 8 9 10—87 86 85 84 83

Library of Congress Cataloging in Publication Data

Levenbach, Hans.
A STATLIB primer.

1. STATLIB (Computer programs) 2. Time-series analysis—Data processing. 3. Business forecasting—Data processing. I. Brelsford, William M. II. Cleary, James P. III. Title. IV. Title: S.T.A.T.L.I.B. primer.
QA276.4.L48 1983   003.2'0285425   82-14068
ISBN 0-534-97936-X

Contents

Preface

This manual is a primer to STATLIB—a statistical computing library. It is also an introduction to the use of the STATLIB system for forecasting applications, particularly time series forecasting.

The objective of the primer is:

- To introduce material describing how to get onto the computer, set up data files for statistical analysis and forecasting, and obtain results
- To provide computer tools for data display, analysis, manipulation, and statistical modeling
- To demonstrate the utility and importance of graphical methods for presenting the results of data analysis
- To illustrate the effectiveness of the software for forecasting time series

STATLIB is a software package designed by practitioners for practitioners. Because it is designed especially for statistical analysts, it can be used in a broad range of business application areas, including:

- Data management, graphical display and information summary
- Sales forecasting and tracking
- Econometric studies
- Model fitting and simulation
- Survey sampling and market segmentation studies

Professional analysts who need to perform a host of vital statistical procedures to support their applications will find this volume invaluable. Chapter 1 describes STATLIB's general capabilities and genesis. The system operates efficiently in an interactive environment under the control of ISX (Interactive STATLIB Executive System), the command language for executing STATLIB routines. The basic concepts of this interactive environment and the getting-started steps for accessing it are described in Chapter 2. Chapter 3 describes various types of information, including data for analysis, temporary results, and ways to manage inputs and outputs efficiently. Further capabilities of the ISX language are treated in Chapter 4.

Treatment of different types of data structures and how they can be analyzed begins with Chapter 5. This chapter describes time series and shows how they can be manipulated, managed and modeled. A different data structure, the data matrix, is handled in Chapter 6. Data matrices, or "variable-by-observation" arrays, are widely used in regression applications, as well as in cross-tabulation, distribution fitting and logit maximum likelihood estimation.

In order to facilitate more complex analyses and procedures, the ISX language allows the user to do some programming. This capability, described in Chapter 7, leads to user defined procedures or macro programs. Chapter 8 summarizes some useful utility routines that are frequently required for terminal session and computer workspace control.

Graphical displays of data, both the inexpensive printer-plot variety and those drawn on high-quality graphics devices, are important in all aspects of data analysis, modeling, and the presentation of results. Chapter 9 is an introduction to STATLIB's graphics capabilities. It shows the user how to develop, modify, and produce graphical output. The capabilities described in this chapter include time plots, bar charts, histograms, pie charts, and box plots.

To complement the primer for forecasting applications, an annotated example is provided as an appendix. It is, we hope, invaluable to the novice forecaster who would like to begin quickly and effectively to understand data, analyze patterns, and develop a forecasting rationale.

Several related documents and books will be of interest to the reader of this primer. The *STATLIB Manual* (Brelsford and Relles, 1981), contains documentation for all the routines and commands in the basic STATLIB system.

STATLIB/ISX System Procedures (Agarwala, Brelsford, Inglis, and Preston, 1982) describes the System Procedures, an extension of STATLIB/ISX containing an extensive set of user-friendly commands written in the ISX language. Most of the commands discussed in this primer are system procedures.

Two other books utilize STATLIB extensively for examples, illustrations, models, and analyses. *The Beginning Forecaster* (Levenbach and Cleary, 1981) and *The Professional Forecaster* (Cleary and Levenbach, 1982) are companion volumes designed for the practicing forecaster. Both volumes stress a process for effective forecasting, with an emphasis on analyzing data before starting model-building. Thus, for forecasters and professional analysts alike, STATLIB offers a comprehensive statistical analysis system for integrated data-oriented analyses.

H. Levenbach
W. M. Brelsford
J. P. Cleary

Acknowledgments

For a number of years (approaching a decade!), the authors have recognized the need for and value of introductory material on STATLIB and its uses, especially for forecasters and econometricians. This primer represents the evolution of the contributions of many capable individuals, only some of whom we can recognize individually for their special efforts.

Foremost, we are grateful to Barbara Wiener, who was responsible for most of the present design, restructuring and rewriting of earlier attempts. Special appreciation is also due to Sherry McArdle Karas, who contributed additional writing (Chapters 6 and 9), editing, preparation of computer examples, and general coordination.

The primer evolved from an early version written by Bruno Korschek. Margaret Rowe made significant advances in the primer's design when she wrote new and revised old material to create the second version. Improvements in these versions benefitted from the comments and suggestions offered by users throughout the Bell System. Their valuable insights are greatly appreciated.

Mary Santucci typed most of the manuscript into the computer and prepared it for the phototypesetter. Joseph McCabe provided invaluable assistance in the preparation of the annotated forecasting example in the appendix. Daniel Relles of The Rand Corporation reviewed the entire manuscript, and made many helpful comments. We also express appreciation to Pramila Agarwala, Thomas Beattie, Robert Brousseau, William Lisowski, Maryann Mercer Fava, David Preston, and Randolph

Reitz for commenting on specific sections and chapters of the final version and for assisting in the development and support of the system.

While we wish to acknowledge our indebtedness to all the contributors, we alone bear responsibility for whatever errors and mistakes are still present in the text.

H. L.
W. M. B.
J. P. C.

CHAPTER 1

Introducing STATLIB/ISX

STATLIB is a software package for statistical data analysis—almost any kind of data analysis, including forecasting of time series. It can be used for almost any type of application requiring the display, summary, manipulation, and modeling of data. Because of its versatility, STATLIB applications have been made in the social sciences, engineering, business, and marketing.

HISTORICAL EVOLUTION

The development of STATLIB began in the mid-1960s at Bell Laboratories, where William Brelsford and Daniel Relles provided statistical computing support and consultation to fellow research statisticians, engineers, and scientists. Over time, as statistical computing gained recognition, the software was built into a cohesive, flexible, and powerful system. The nature of the work at Bell Labs and the needs of the researchers determined STATLIB's design criteria. Among these, flexibility of use and the integrity of the routines have contributed significantly to STATLIB's wide acceptance and use within the Bell System, especially among forecasters and econometricians.

With the publication of the *STATLIB Manual* (Brelsford and Relles, 1981), the system has become more widely known in other user communities. This primer thus serves as an introduction for all users to the STATLIB system. Moreover, because of the annotated example, it is especially well-suited for forecasters and econometricians.

STATISTICAL ANALYSIS ROUTINES

STATLIB is a flexible, interactive system for performing analyses of data. It consists of an extensive library of FORTRAN-callable subroutines and an interactive command language, ISX, which permits easy on-line access to the subroutines.

STATLIB's capabilities are wide and far-reaching. Its functions range from basic operations, such as data display, tabulation, and elementary algebraic functions, to sophisticated statistical analyses, such as *regression*, *spectrum analysis*, and *econometric modeling*.

The ISX language provides a powerful tool for *interactive analysis*. With it, users can not only access routines in the FORTRAN library but also build their own commands. ISX allows for arbitrary user-written commands, indistinguishable from other commands. This capability is a strong feature of the system since it provides branching, conditional execution, parameter passing, nested calling (including recursion), and so forth. It facilitates development of applications ranging from turnkey systems for non-statisticians to sophisticated statistical procedures built up from available routines. In addition, STATLIB provides a convenient framework of data management and storage allocation routines to which new FORTRAN-callable subroutines can easily be added.

STATLIB's graphics package, GRPAK, generates high-resolution graphics output on a variety of output devices (CRT terminals, flatbed plotters, microfilm, etc.). Capabilities include scatter plots, time plots, bar charts, pie charts, box plots, "bubble" charts, and maps.

STATLIB operates on three principal data structures: *time series* (observations on a variable at different times), *two-way data matrices* (a set of observations on a group of variables, as used in regression), and *multidimensional tables* (e.g., number of employees classified by race, sex, level, age, etc.). Data may be maintained in one structure and transformed easily to the others.

In a totally interactive, command-driven environment, TSPAK, STATLIB's time series package, provides the tools to produce credible and timely analyses, forecasts, and displays of time series data. Market forecasters, corporate planners, econometricians, and statistical analysts can use TSPAK to develop and present immediate results that are responsive to management demands for quantitative information. Software tools are available for applications that include:

- Seasonal and trend analysis
- Exponential smoothing
- Lead-lag analysis
- Periodicity conversion (month-quarter-year etc.)
- Residual analysis
- Autocorrelation correction
- Distributed lags
- Two-stage least squares
- Univariate Box-Jenkins modeling
- Forecast testing

Communicating important information in tables and graphs saves time and words. TSPAK, through its interface with STATLIB's graphics package, GRPAK, offers a wide variety of graphical output, from simple printer-plots to high-quality, high-resolution displays. Some of the features include:

- Autocorrelation plots
- Scatter plots
- Time plots
- Residual plots

STATLIB's statistical analysis package for two-way data matrices, STPAK, provides many of the same benefits as the time series package, TSPAK. Using STPAK, an analyst can efficiently obtain many summary statistics (including *robust/resistant* measures), perform statistical analyses, and obtain a wide variety of plots.

As with TSPAK, STPAK is easy to use and provides ready access to STATLIB's programming capabilities and to other packages, such as the graphics package, GRPAK.

Analysts use STPAK to perform many kinds of data analysis on two-way data matrices. Some of the applications include:

- Analysis of survey data
- Marketing analysis
- Medical data analysis
- Analysis of census data
- Demand analysis
- Financial analysis

STPAK procedures offer the user a wide variety of analytical tools, ranging from traditional ordinary least squares regression to innovative procedures such as *box plots* and *robust regression* techniques. Some of the analysis tools to choose from are:

- Descriptive statistics
- Cross tabulation
- Simple and multiple regression
- Stepwise regression
- Robust regression
- Ridge regression
- Generalized least squares
- Pooling
- Logit maximum-likelihood estimation
- Cluster analysis
- Distribution fitting
- Random sampling

STPAK can be used to interpret data and arrive at management decisions. Many times the rationale for such decisions must be presented in a timely, easy-to-interpret graphical fashion. As with TSPAK, STPAK has an interface to GRPAK, STATLIB's graphics package.

STATLIB's third principal data structure is the multidimensional table. Routines in TABLE operate on data of this type. Capabilities include data manipulation, such as projection, extraction, and transposition, and basic modeling. The TABLE package, similar in design to TSPAK and STPAK but inherently more complex, is not discussed in this primer.

CHAPTER 2

Getting Started

ISX—the Interactive STATLIB Executive System—makes STATLIB's analytical capabilities available to users in a convenient, easy-to-use, interactive environment.

A general understanding of the interactive environment in which ISX operates is essential to the effective use of the system. This chapter begins by describing the interactive operating system, including procedures for accessing the computer, storing information, and invoking ISX.

The chapter then discusses basic concepts that apply to the ISX environment in general, including session control, data structures, and ISX grammar and notation conventions.

THE INTERACTIVE ENVIRONMENT

ISX is accessed under the control of an interactive computer *operating system*. The operating system permits multiple users to share the resources of a large-scale host computer system.

ISX may be installed on various hardware under the control of various operating systems, but it is most commonly available on systems using IBM's VM/CMS, which stands for Virtual Machine/Conversational

Monitor System. Under VM/CMS, each use has access to a dedicated portion of the computer, a *virtual machine*, independent of other users on the system. A user's virtual machine is illustrated in Figure 2.1.

The user converses with the computer using a typewriter-like device called a terminal, which communicates with the computer via telephone lines. This conversation, or interactive dialogue, consists of the user typing *commands* at the terminal keyboard that the computer operating system recognizes and processes. The computer then transmits responses that are received and displayed at the terminal. Finally the computer indicates that it is ready to receive and process the next command.

Beginning a Terminal Session

To gain access to the computer system, you will need the appropriate telephone number and an assigned *userid* and *password*. This information is usually provided by the computer system coordinator or administrator responsible for your installation.

You dial the phone number and, upon hearing the high-pitched tone, place the phone receiver in the terminal's acoustic coupler (or press the data button on the telephone or terminal). Specific *log-on procedures* may vary depending on your operating system—on VM/CMS, depress the *carriage return* key, and wait for the computer response:

```
VM/370 ONLINE
```

The "." indicates that the computer system is ready for your response. (The use of *prompt characters* may vary on some systems.) Enter the LOGON command (followed by a carriage return):

```
LOGON userid password initials
```

For example, if your *userid* is myuser, your *password* secret, and your initials brw:

```
logon myuser secret brw
```

The system will verify that you have entered a valid *userid* and *password* combination, print the date, time, and possibly other information, and indicate that it is ready to receive a command:

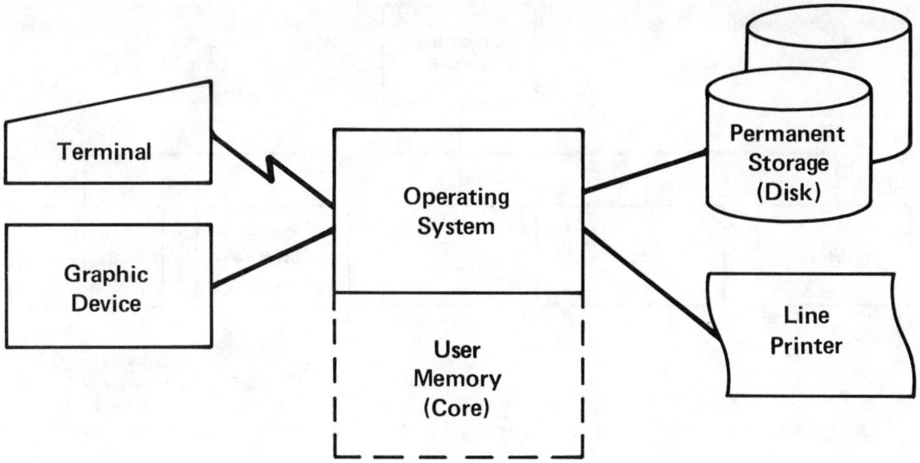

Figure 2.1. Each user has access to a dedicated virtual machine.

```
LOGON AT 08:15:45 NYT MONDAY 03/23/82
CMS VERSION 82.1
R;

.
```

You now have access to all functions provided by the operating system. You can enter any valid system command, followed by a carriage return. From the operating system level you can call the system editor to create and modify files:

```
. edit
```

You can invoke ISX to perform statistical analysis:

```
. isx
```

You can also issue system commands that manipulate existing files, and execute your own programs or other existing software packages.

The relationship of various environments to the operating system is illustrated by Figure 2.2. Each environment (e.g., editor, ISX, etc.) has its own set of commands.

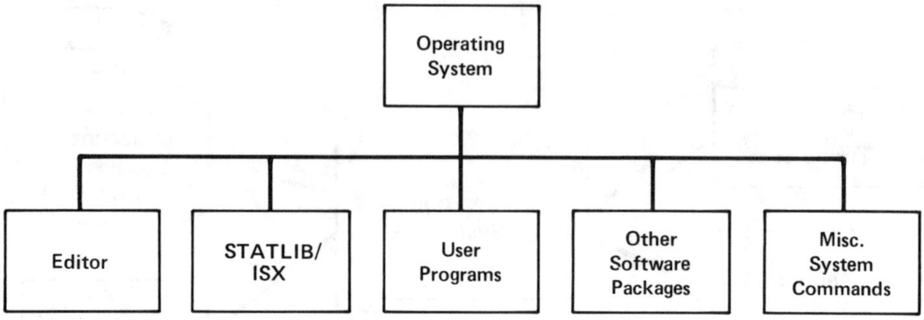

Figure 2.2. Computer environments as related to the operating system.

Ending a Terminal Session

To end a terminal session from the operating system level, use the system's LOGOFF command; on VM/CMS:

```
. logoff
```

LOGOFF prints session summary information and disconnects the terminal from the computer.

PERMANENT STORAGE—THE FILE SYSTEM

Files provide permanent storage for information that is to be saved for later use or made available to other environments within the operating system. A file containing data may be created using the editor for subsequent retrieval by ISX. Results generated in ISX may be saved in a file for later listing on a high-speed printer or for plotting on a graphics device. Files may contain a set of data (a *data file*), a set of computer instructions (a *program file*), a list of system commands (a *command file*), or program-generated results (an *output file*). Once a file is created, it usually remains available until it is erased (deleted).

The operating system keeps track of all files related to a particular *userid*. System commands are available to manipulate existing files, for example, to print, copy, rename, or erase files.

On VM/CMS, files are stored on *virtual disks*. The so-called A-disk is the user's primary read/write disk; it is automatically available at the beginning of a terminal session. Additional disks may be made available as required.

Naming Files

A file is identified by a *file identifier* or *fileid*. On VM/CMS, a fileid consists of three fields, called *filename*, *filetype*, and *filemode*. The filename and filetype may each contain one to eight *alphanumeric* characters (any combination of the letters A through Z and numbers 0 through 9). The filename identifies the contents of the file—For example:

MYDATA
FORECAST
PROCPLOT

The filetype usually denotes the type of information contained in the file—For example:

DATA	to denote a data file
ISX	to denote an ISX command file

The filemode specifies the virtual disk on which the file resides (A, B, C, etc.) If not specified, the system usually defaults to the A-disk or searches each disk in sequence.

Some examples of complete VM/CMS fileids are:

MYDATA DATA A
FORECAST DATA A
PROCPLOT ISX C

On other operating systems, file naming conventions may vary. If such is the case, the VM/CMS fileids used in examples throughout this primer should be replaced following the proper convention.

Creating a File

Files containing data or commands to be used with ISX may be created or modified within the editor. On VM/CMS, the editor has two modes: the INPUT mode for creating new files (or inserting additional information in existing files), and the EDIT mode for modifying files.

To create a new file with fileid "FORECAST DATA A," invoke the CMS editor by entering the command:

```
. edit forecast data
```

Since the filemode was not specified, it defaults to A. Since FORECAST DATA A doesn't already exist, the system will respond with:

```
NEW FILE:
EDIT:
.
```

You are now in the EDIT mode, with an empty *temporary file* or *workspace*. To insert data into this temporary file, enter the INPUT mode by typing the `INPUT` command:

```
. input
INPUT:
.
```

From this point on, each line of information entered adds a line to the temporary copy of the file. To terminate data input, enter a *null line*, that is, a line containing nothing but a carriage return:

```
. 1000
. 1500
. 2000
. 2500
. 3000
. 3500
. 4000
.            ← null line terminates input
EDIT:
.
```

The terminating carriage return exits the INPUT mode and returns you to the EDIT mode. At this point you may type out some or all of the temporary file, or make editing modifications, such as correcting errors by replacing a string of characters, and replacing or deleting lines.

Assuming that the file has been input correctly and that no corrections are required, the current temporary copy of the file may now be stored as a *permanent file* in order to be available to the system. To save the file and exit the editor, type the **FILE** command:

```
. file
R;

.
```

The operating system will respond with the prompt character indicating that it has exited from the editor and is ready to accept a system command.

To summarize the procedure for creating a new file:

```
. edit filename filetype [filemode]
NEW FILE:
EDIT:
. input
INPUT:
.    (type each line of data)
.    .
.    .
.    .
.
EDIT:
. file
R;

.
```

Like the **FILE** command, the **SAVE** command stores the current temporary file copy as a permanent file, but **SAVE** leaves the computer in the editor. The **QUIT** command exits the editor without saving the current temporary file copy.

TEMPORARY WORKSPACE—MEMORY

While files provide a medium for permanent storage of information external to the computer's central processor, memory (or core) provides temporary internal workspace for processing information and storing temporary data and results. For example, in the CMS editor described in the previous section, the temporary file copy being input or edited is stored and manipulated in memory. It doesn't become permanently recorded for subsequent use until it is `FILE`d or `SAVE`d.

When beginning a terminal session, a minimum amount of memory is automatically allocated. On VM/CMS, the default size of memory automatically allocated to each user is 320K. K is a unit of measurement of computer storage capacity equal to 1024 characters or *bytes*.

In order to use ISX, the amount of memory allocated must be large enough to accommodate both the ISX system routines and a storage area for data, procedures, and temporary results. ISX uses the allocated space as illustrated in Figure 2.3. A portion of the space is occupied by the actual STATLIB/ISX routines. The remaining space is available to store data for analysis, temporary variables and results, and ISX and user-defined procedures.

The memory requirements necessary for ISX must be defined at the operating system level prior to invoking ISX. On VM/CMS, the `DEFINE STORAGE` system command can be used to redefine the amount of memory available. For example, to allocate 800K:

```
. define storage 800k
```

The amount of memory allocated can also be defined initially via the `LOGON` command by adding the phrase "CORE nK"—For example:

```
. logon myuser secret brw core 800k
```

Note that memory requirements and the commands used to allocate memory may vary greatly from one computer system to another.

ACCESSING ISX

In addition to allocating sufficient memory to use ISX, the program files containing the STATLIB/ISX system routines must be made available prior to invoking ISX. On VM/CMS this is accomplished by "attaching" the disk on which the STATLIB/ISX routines reside. In addition to the A-disk, the primary read/write disk automatically attached at log-on, additional disks (B, C, etc.) can be attached as needed. In order to attach the STATLIB disk, enter the following system command:

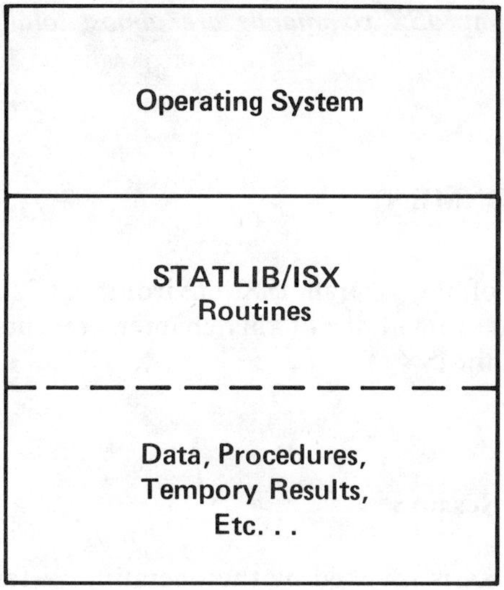

Figure 2.3. Space in the user's virtual machine is allocated to the operating system, to ISX, and to user data.

```
. attach statlib btl rr
```

In this case, "statlib" is the *userid*, "btl" is the *password* associated with the disk, and "rr" specifies read-only mode. The system will respond by indicating that it has successfully attached the specified disk, and indicate the filemode (B, C, etc.) assigned.

You are now ready to invoke ISX to perform statistical analysis of your data. To enter the ISX environment, type the command:

```
. isx
```

The following list summarizes the VM/CMS system commands required to access ISX:

```
. logon userid password initials
. define storage 800K
. attach statlib btl rr
. isx
```

Note that these commands apply on VM/CMS. On other operating systems, the required system commands may vary. However, once within the ISX environment, *ISX commands are almost totally independent of the operating system.*

THE ISX ENVIRONMENT

An understanding of the overall ISX environment is necessary for using ISX effectively. The remainder of this chapter presents basic concepts and rules that apply within ISX.

An ISX Terminal Session

As we have seen, ISX is invoked by the operating system—level command:

```
. isx
```

Once within ISX, only ISX commands may be entered. The environment is controlled by a set of defaults or system options. The first command that should be entered is:

```
. term
```

`TERM` performs some required initialization functions and establishes various system options appropriate for interactive processing from a terminal. Subsequent commands can be entered to retrieve or create data, perform the desired statistical analysis (regressions, correlations, forecasting, etc.) on the data, and generate results in printed and/or graphical form.

To end the session, enter the ISX statement:

```
. /log
```

to exit from ISX and log off the system, or enter the ISX statement:

```
. /stop
```

to exit ISX and return control to the operating system level.

Data in ISX

ISX is normally used to analyze two types of data: time series and two-way matrix. (A third type, the multidimensional matrix, or table, is beyond the scope of this primer.)

A *time series* is a collection of ordered observations over time (or space). In addition to the actual data values, a time series has an associated start date and a two-level time scale, for example, years and months, years and quarters, days and hours, and so on. Some examples of data sets that might be represented as time series are:

Housing Starts by month and year
Money Market Interest Rates by quarter and year
Total Inches of Rainfall by day and month

A *two-way matrix* is a two-dimensional data structure in which one dimension represents cases or *observations*, and the other dimension represents *variables* or measurements on these observations. A simple example of a two-way matrix contains three variables—income, household size, and number of telephones—for twenty-three households (observations).

Data must reside in the ISX core workspace before analytical routines can process them. The data can be retrieved from existing data files or entered directly from the terminal, as shown in Figure 2.4.

Once the data reside in the ISX core workspace, the type of data determines the analysis routines that can be applied to them. Recall from Chapter 1 that one set of routines (TSPAK) is available for time series analysis and another set (STPAK) is provided for two-way matrix analysis, as shown in Figure 2.5.

ISX Statements

ISX statements are entered one at a time and processed immediately. (Exceptions, including ISX command files and user-defined procedures, are discussed in Chapter 7.) The following general rules apply:

- Statements consist of a command verb, possibly followed by one or more arguments.
- All arguments and required punctuation must be separated by one or more blanks.

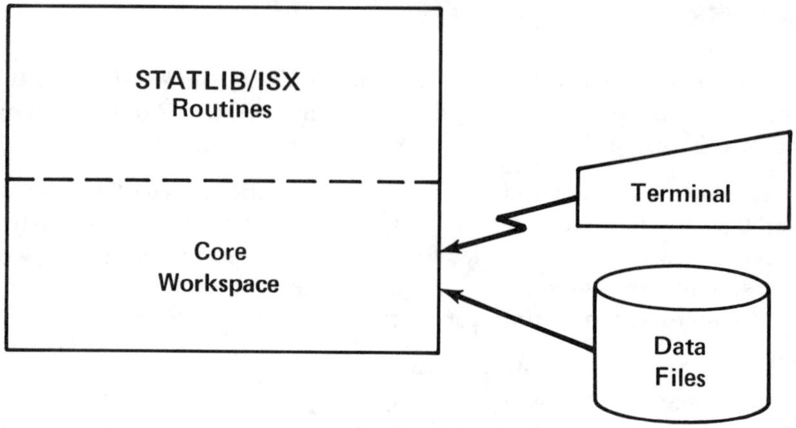

Figure 2.4. Data can be retrieved from existing data files or entered directly from the terminal.

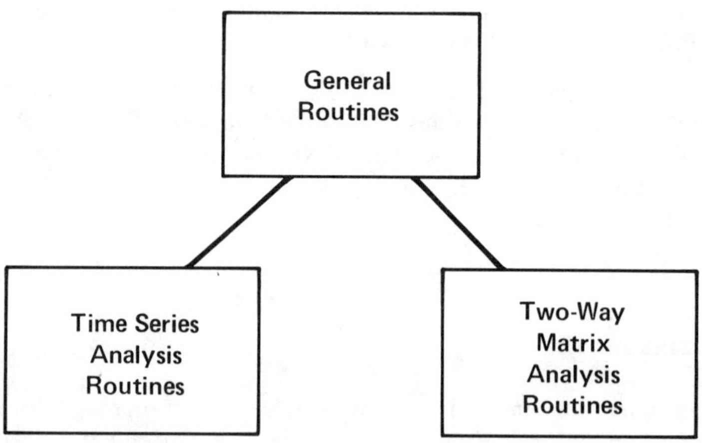

Figure 2.5. Routines are available for analyzing time series and two-way matrix data, among others.

- Statements may begin at any point on a line.
- Each line may contain up to 80 characters.
- A statement may be continued on the succeeding line by including the character "$" as the last argument on the current line.

Statement Notation

Where general ISX statement formats are included in this text, the following notation will be used:

- Items appearing in upper case are command verbs and keyword arguments that must be entered as shown, unless otherwise specified.
- Items appearing in lower case indicate information to be supplied, such as file or data identifiers.
- Items included in square brackets ([ ]) are optional.
- Where applicable, alternative optional arguments are separated by a vertical bar (|).
- All punctuation appearing is required and must be entered as shown.
- ".." indicates that the preceding item(s) may be repeated as needed.

Prompt Characters

As noted previously, the computer system transmits a *prompt* character to indicate its readiness to accept terminal input. In order to distinguish examples of commands within the ISX environment from operating system–level commands, we have adopted the convention of indicating the prompt character ">" when within ISX and the prompt character "." in all other situations. In reality, the prompt character in both cases is identical and depends upon the system default or user specification. (On systems not using prompt characters, a line feed indicates that the system is ready to receive the next command.)

Error Messages

Common ISX errors include misspelling of a command verb or argument keyword, leaving out or misspelling a data identifier, and having insufficient core (memory) available. A command not recognized by ISX will result in the simple message "?." Statements that cannot be processed will result in an appropriate error message. An erroneous data identifier (time series, matrix, variable, etc.) will result in the message "blocktype blockname NOT FOUND," meaning that the data referred to do not exist. If insufficient core storage remains to create a new data item, the message "CORE NOT AVAILABLE FOR blocktype blockname" will result.

Execution Messages

Many ISX statements execute STATLIB routines, either explicitly or implicitly. Most STATLIB routines generate a one- or two-line message summarizing what they have done. This is useful for tracing the flow of execution through the STATLIB routines and results in making output self-documenting.

Message Control

The `TERM` command, which we used to initialize ISX system parameters for interactive processing, permits a virtually unlimited number of errors to occur during the interactive session. The `OPTION` statement, discussed in Chapter 8, may be used to limit the number of errors permitted or to suppress printing of error messages and execution messages.

CHAPTER 3

Data in ISX

As we have seen in the previous chapter, ISX provides a temporary workspace area in core for use during an ISX session. This area is used to accommodate various types of information, including data for analysis, temporary results, ISX variables, predefined macro routines (called *procedures*), and ISX control information. Capabilities are provided to dynamically allocate and manage this area of core. Information residing in this "dynamic core" area is organized in blocks, referred to as DYNCOR blocks.

DYNAMIC CORE

Data must reside in core in DYNCOR *blocks* before they can be referenced in subsequent ISX commands, analyses, and calculations. Data sets may be created or brought into core in several ways, as illustrated in Figure 3.1. Data may be retrieved from existing permanent files, entered interactively from the terminal, or possibly calculated as a function of other data or as a result of statistical analysis.

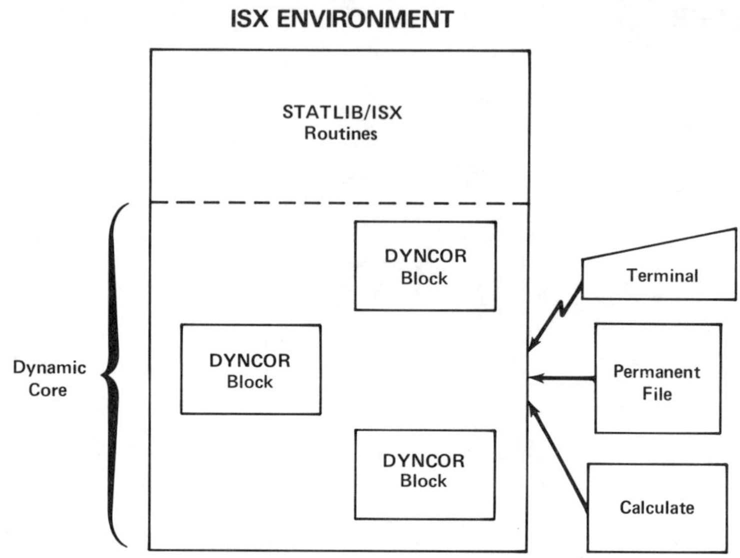

Figure 3.1. Data sets may be created or brought into core in several ways.

DYNCOR Block Identifiers

Once a data set resides in dynamic core, it can be accessed or referenced by its designated *block identifier.* A block identifier consists of two parts: a *blocktype* and a *blockname.* The blocktype and the blockname can each consist of up to four alphanumeric characters. Blocktypes beginning with punctuation marks have special meaning in ISX and usually indicate that the block contains *control information.* Typical block identifiers are illustrated in the following table:

Block-type	Block-name	Description
DATA	REV	Revenue time series data
PRDN	REV	Revenue predictions or forecasts
RSID	REV	Revenue residuals
DATA	MTRX	Two-way matrix MTRX
*DAT	MTRX	Control information for two-way matrix

For purposes of clarity, block identifiers appearing in this text will be enclosed in angle brackets, i.e., <type name>. Remember, however, that the brackets are not part of the block identifier and should not be included in actual ISX commands.

TIME SERIES DATA

A time series in ISX consists of a data set of ordered observations and associated time control information. The most common method of entering time series data into ISX for subsequent analysis is to retrieve them from an existing permanent file. Consider the permanent file FORECAST DATA A, whose creation in the editor was illustrated in Chapter 2. The VM/CMS `TYPE` command prints the contents of this file as follows:

```
. type forecast data

1000
1500
2000
2500
3000
3500
4000
```

Let us assume that the data set contained in this file represents a time series of annual revenue data. In order for this data to be accessible to ISX, additional information must be provided: A block identifier must be assigned to this data set, and the number of items to be input, the data type of the items (i.e., *real* or *integer*), and a format indicating how the system should read in the data must be indicated. This information can most conveniently be specified by adding a line to the file preceding the data, such as:

```
DATA REV 7 R(*)
```

This line, called a *header record*, indicates that seven items of real (i.e., floating-point) data are to be read using format "(*)," meaning *free format*. Free format input allows one or more items per line, separated from each other by at least one space. Now the file appears as follows:

```
DATA REV 7 R(*)
1000
1500
2000
2500
3000
3500
4000
```

The file now contains sufficient information for ISX to retrieve this data set as DYNCOR block <DATA REV>. However, our time series specification is still incomplete. Another block, called a *time control block*, must be provided to define the start date and periodicity associated with the time series data contained in <DATA REV>.

Time Control Blocks

A time control block also has a two-word block identifier. Its blocktype consists of a period (.) followed by the first three characters of the blocktype of the associated data block. Its blockname is identical to the blockname of the associated data block. Thus, time control block <.DAT REV> provides the start date and periodicity associated with the time series data contained in <DATA REV>.

The time control block must provide three items of information to specify the date structure of an associated time series: a *start-year*, *start-month* and number of months per year. Note that "year" and "month" are generic terms representing any two-level time scale, where "year" represents the major time period, "month" the minor time period and "months per year" the number of minor time periods per major time period. A one-level scale, annual data for example, can be represented as having one month (minor time period) per year.

If the time series <DATA REV> contains annual data beginning in 1975, the time control information required to specify this date structure would be as follows:

start-year = 1975
start-month = 1
months-per-year = 1

Thus, the time control block <.DAT REV> could be added to the file by inserting two lines, a header record:

```
.DAT REV 3 I(*)
```

indicating that <.DAT REV> contains 3 integer items to be read using free format, and a line containing the appropriate time control information:

```
1975 1 1
```

Now the file appears as follows:

```
.DAT REV 3 I(*)
1975 1 1
DATA REV 7 R(*)
1000
1500
2000
2500
3000
3500
4000
```

It contains two blocks: a data block, called <DATA REV>, and its associated time control block, <.DAT REV>. Both blocks are required to specify the time series referred to in ISX as <DATA REV>.

Header Records

As we have seen, each block, whether containing data or control information, is preceded in an input data file by a header record having the following general structure:

blocktype blockname len t(format)

where:

len = number of items to follow

t = data type of items to follow
(r for real, i for integer)

format = format to be used to read data

The format may be a standard FORTRAN format statement (e.g., 4F8.3), or "*," specifying that the data items to follow are to be read in free format, that is, one or more items per line, separated by blanks.

Retrieving Time Series from a File

It is essential to understand that up to this point the data exist on a permanent file external to the ISX environment. Files are created, modified, and saved via the editor prior to invoking ISX. Once the permanent file FORECAST DATA A exists, copies of all blocks contained in it can be brought into the ISX environment by issuing the ISX `INPUT` statement. For example:

```
> input forecast
INPUT FROM FORECAST DATA ...
  DYNINP 28:   .DAT REV
  DYNINP 28:   DATA REV
```

Note that only the *filename* need be specified. On CMS, ISX defaults the *filetype* to DATA and the *filemode* to A. The message "DYNINP 28: blocktype blockname" is an execution message generated by the STATLIB routine `DYNINP`, which has been implicitly executed by the `INPUT` statement. The message provides on-line verification of each block that has been copied from the permanent file into dynamic core.

A permanent file may contain any number of blocks. If the file FORECAST DATA A also contains another time series, <DATA INDX>, representing a monthly series beginning in May 1978, the file might appear as follows:

```
.DAT REV 3 I(*)
1975 1 1
DATA REV 7 R(*)
1000
1500
2000
2500
3000
3500
4000
.DAT INDX 3 I(*)
1978 5 12
DATA INDX 23 R(*)
1.34 1.39 1.42
1.43 1.41 1.59 1.67 1.73
1.62 1.58 1.61 1.623
1.67 1.72 1.72 1.79 1.81 1.83
1.79 1.791 1.83 1.91 1.9
```

Now the `INPUT` statement would retrieve all four blocks contained in the permanent file, and store copies of them as DYNCOR blocks within the ISX environment:

```
> input forecast
INPUT FROM FORECAST DATA ...
  DYNINP 28:  .DAT REV
  DYNINP 28:  DATA REV
  DYNINP 28:  .DAT INDX
  DYNINP 28:  DATA INDX
```

The general structure of a permanent data file and the resulting DYNCOR blocks retrieved within ISX are illustrated in Figure 3.2.

The `INPUT` statement can also be used for selective retrieval of a subset of blocks contained in a file by specifying the list of blocks to be retrieved:

```
> input forecast data rev data msg
INPUT FROM FORECAST DATA ...
  DYNINP 28:  .DAT REV
  DYNINP 28:  DATA REV
```

Notice that control blocks (here <.DAT REV>) are loaded automatically along with data blocks, and that blocks not found in the file (such as <DATA MSG>) are ignored quietly.

A shorthand notation can be used to specify the retrieval of blocks with specified blocknames but arbitrary blocktypes. For example, we can specify the retrieval of only blocks with blockname REV; this is illustrated by the following:

```
> input forecast .... rev
INPUT FROM FORECAST DATA ...
  DYNINP 28:  .DAT REV
  DYNINP 28:  DATA REV
```

The string "...." serves as a *global argument* in ISX. In this case, the command specifies retrieval of all blocks having a blockname of REV regardless of the blocktype. The global argument "...." can be used in either the blocktype or blockname position.

Entering Time Series

In addition to retrieving time series from permanent files, new time series may be entered interactively from the terminal using the `ENTER` statement—For example:

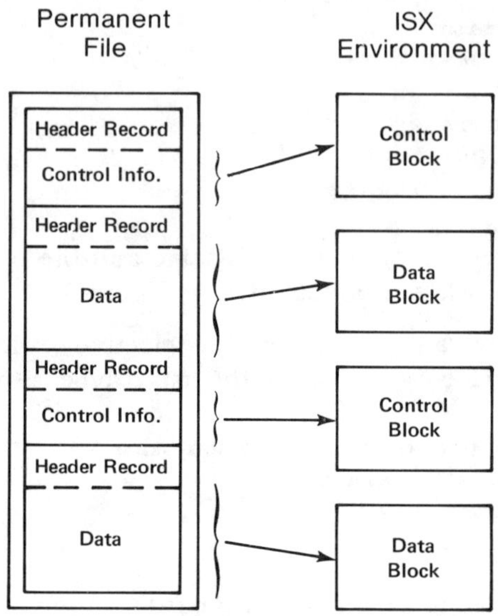

Figure 3.2. A permanent data file may contain any number of DYNCOR blocks.

```
> enter data rev torg 1975 1 1 $
> 1000 1500 2000 2500 3000 3500 4000
```

In this example, "data rev" is the series to be created, "torg 1975 1 1" specifies the time origin (start-year, start-month, number of months per year) and "1000 1500 2000..." are the actual data values. (Note the use of the continuation argument, $, to continue input of a command on the next line.)

We can verify that the two DYNCOR blocks comprising the series <DATA REV> have actually been created; we do this by using the /LIST command, requesting that it list all DYNCOR blocks having a blockname of REV:

```
> /list .... rev
DATA REV
.DAT REV
```

We can also use the **NAMETS** command to verify creation of the time series <DATA REV> and its associated time origin:

```
> namets t data rev
DATA REV   1975-1981
```

Calculating Time Series

Using the `SET` command, new time series can be created in ISX as arithmetic functions of existing series. For example:

```
> set data lrev = log ( data rev )
```

creates a new series <DATA LREV> consisting of the logarithms of the series <DATA REV>. The values of both series can be printed using the `PRTTS` command:

```
> prtts time 1975 1 .1981 1 data rev data lrev

OBL          DATAREV  DATALREV

1975         1000.00      6.91
1976         1500.00      7.31
1977         2000.00      7.60
1978         2500.00      7.82
1979         3000.00      8.01
1980         3500.00      8.16
1981         4000.00      8.29
```

Saving Time Series

It is important to understand that any new time series created during an ISX session exist only within the current session and will no longer be available once ISX is exited. The `OUTPUT` statement can be used to record copies of current time series on a file for permanent storage. The file can then be accessed in subsequent ISX sessions using the `INPUT` statement described previously.

Assume, for example, that the entire contents of the file FORECAST DATA A have been input into ISX, some analysis has been performed, and three new series (and their associated time control blocks) have been created:

<DATA LREV> and <.DAT LREV>
<DATA NEW> and <.DAT NEW>
<PRDN REV> and <.PRD REV>

To save these series on a new permanent file called NEWFILE DATA A, the OUTPUT statement is used. For example:

```
> output newfile data lrev data new prdn rev
OUTPUT (APPEND) TO NEWFILE DATA ...
  DYNOUT: .DAT LREV 28
  DYNOUT: DATA LREV 28
  DYNOUT: .DAT NEW  28
  DYNOUT: DATA NEW  28
  DYNOUT: .PRD REV  28
  DYNOUT: PRDN REV  28
```

Since the file NEWFILE DATA A does not already exist, it is created and now contains six DYNCOR blocks: three time series data blocks and three associated time control blocks. It is not necessary to specify the time control blocks in the OUTPUT command sequence. OUTPUT stores them in the file automatically. Note that only the *filename* is specified. On CMS, the *filetype* defaults to DATA and the *filemode* to A.

The new series can also be stored in the original file:

```
> output replace format auto forecast $
>   data lrev data new prdn rev
OUTPUT (REPLACE) TO FORECAST DATA ...
  DYNOUT: .DAT LREV 28
  DYNOUT: DATA LREV 28
  DYNOUT: .DAT NEW  28
  DYNOUT: DATA NEW  28
  DYNOUT: .PRD REV  28
  DYNOUT: PRDN REV  28
```

The REPLACE keyword instructs OUTPUT to replace existing data, if any, with the current data. Otherwise the new data are appended to the end of the existing file.

The FORMAT AUTO option selects an appropriate output format to accommodate the data. If not specified, the file is generated in an internal code, which cannot readily be printed or edited, but is more accurate and can be read more efficiently.

More about Time Series

Chapter 5, on Time Series Analysis and Forecasting, provides a more complete discussion of time series data management.

TWO-WAY MATRIX DATA

A two-way matrix is another type of data set that is commonly processed in ISX. A two-way matrix is a two dimensional data structure, where one dimension represents cases or *observations* and the other dimension represents *variables* or measurements on these observations.

The sample matrix that follows represents responses to a questionnaire. This matrix contains three variables: AGE, SEX, and INCOME. Each variable contains five observations, where each observation corresponds to a person's responses.

	Variables		
	AGE	SEX	INCOME
	18	0	19.1
	28	0	24.1
Observations	34	1	39.4
	22	1	27.1
	17	0	8.7

A matrix in ISX has a name of up to four alphanumeric characters. Just as a time series requires two DYNCOR blocks in ISX, a two-way matrix is similarly defined by a data block and a block containing a list of variable names.

The data block contains the actual data values for all variables for each observation. It always has a blocktype of DATA. The companion block, containing the list of variable names included in the matrix, has a blocktype of *DAT. The blockname of both blocks is the same as the matrix name. If the sample matrix above is named QUES, then the two DYNCOR blocks that define it are:

<DATA QUES> data block
<*DAT QUES> list of variable names

Within ISX, some sets of statements require specification of the matrix name (e.g., QUES) while other statements require its block designation (e.g., DATA QUES).

An additional variable within a matrix, called "OBL," may be used to contain observation labels. This variable receives special treatment in ISX since it contains character rather than numeric data. Adding the OBL variable to matrix QUES will have the following result:

	Variables			
	OBL	AGE	SEX	INCOME
	Lee	18	0	19.1
	Pat	28	0	24.1
Observations	Terry	34	1	39.4
	C.J.	22	1	27.1
	Chris	17	0	8.7

Like the time series data described previously, two-way matrices can also be retrieved from external permanent files using the `INPUT` statement, and output to files using the `OUTPUT` statement. Files containing two-way matrices follow the same general structure previously described. Each matrix requires a data block and a control block. Each block contains a header record followed by its associated information.

Similarly, commands are provided to enter matrix data interactively, to list matrix and variable names, and to print out matrix data. Chapter 6, on Two-Way Matrix Analysis, provides a more complete discussion of matrix data management.

TIME SERIES VERSUS MATRIX DATA STRUCTURES

The actual data and the analysis required usually determine the appropriate data structure. ISX provides commands to convert a set of time series to a two-way matrix (`TSTOST`), and conversely, to convert matrix variables to time series (`STTOTS`).

ISX VARIABLES

An *ISX variable* is a special data item that may contain any literal string or value. When an ISX variable appears in an ISX statement, it is automatically replaced by its assigned literal string or value. ISX variables are useful in minimizing the amount of typing required when a particular string is likely to appear repeatedly in subsequent commands. ISX variable names begin with a $, followed by one to three alphanumeric characters, the first of which cannot be numeric. Examples are $TIM, $EX2, and $A.

Defining Literal ISX Variables

An ISX variable containing a literal string can be created by the `/DEFINE` statement.

```
/DEFINE $var string
```

For example:

```
> /def $t 1970 1 1980 12
```

Now all subsequent appearances of $T in an ISX statement will be replaced by the literal string "1970 1 1980 12." (The use of $T to specify a date range has special implications in time series analysis and forecasting, which are discussed in Chapter 5.)

A string specification can itself include an ISX variable:

```
> /def $x time $t
```

In this case, all subsequent appearances of $X will be replaced by the string "TIME 1970 1 1980 12."

In the following example, a list of time series appearing repeatedly in a set of commands is replaced by the ISX variable $SER:

```
> /def $ser data rev data msg data hous
> prtts $ser
  (...prints each series in $SER)
> acplt $ser
  (...generates autocorrelation plots of each series in $SER)
```

Defining Integer ISX Variables

An ISX variable may also be defined as the *integer* result of an arithmetic operation: addition(+), subtraction(−), multiplication(*), or division(/), using the `/DEFAI` statement:

```
/DEFAI $var value +|-|*|/ value
```

as illustrated by the following examples:

```
> /def $X 5.1                       $X=5.1
> /defai $X2 2 * $X                 $X2=10
> /defai $sum 8 + 9.7               $SUM=17
> /defai $quo $X / $X2              $QUO=0
```

Note that results of integer operations are always truncated to integer values.

Defining Real ISX Variables

An ISX variable may also be defined as the result of a *real* arithmetic operation, using the /DEFAR command:

```
/DEFAR $var function [ arg1 [ arg2 ] ]
```

as illustrated by the following examples:

```
> /def $X 5.1                       $X=5.1
> /defar $sum add $X 13.2           $SUM=18.3
> /defar $dif sub 92.5 $X           $DIF=87.4
> /defar $max max $X 100            $MAX=100
```

See the *STATLIB Manual*, Chapter 4, for a complete list of available functions.

Deleting ISX Variables

To delete an ISX variable, issue the /DEFINE statement without including a value to be assigned. For example:

```
> /def $t
```

deletes the variable $T.

Listing ISX Variables

The command / (slash) can be used to print the contents of an ISX variable. Following substitution for ISX variables, the command simply prints its arguments, as illustrated by the following examples:

```
> / $t
1970 1 1980 12
> /def $x 3.1
> /def $y 10.0
> / the value of x is $x
THE VALUE OF X IS 3.1
> / x = $x .. y = $y
X = 3.1 .. Y = 10.0
```

PROCEDURES

Another kind of information that may reside in dynamic core is a predefined macro routine called a *procedure*. Procedures are described in Chapter 7, which describes Programming in ISX.

SAVING AND RESTORING AN ISX SESSION

The `/SAVE` command can be used to save all information currently in dynamic core, including all time series, matrices, temporary results, procedures, ISX variables, and ISX control information, in a file for recovery at a later time. Once saved, the contents of this file can be restored via the `/RESTORE` command, recovering the status of dynamic core as it was at the time `/SAVE` was executed.

This capability may be useful if the ISX session must be interrupted for some reason and resumed at a later time—For example:

```
. isx
> term
    (...various ISX commands)
> /save
> /stop
    (...log off or perform non-ISX tasks)
. isx
> term
> /restore
    (resume previous session)
```

`/SAVE` and `/RESTORE` may also be useful within an ISX session, when a branch in the analysis is taken at a particular point. You can save the current session status, proceed with your analysis, and then return to the session status as it was before the branch was taken. For example:

```
> /save
    (...branch in analysis)
> /restore
    (return to previous status)
```

On CMS, if `TERM` has been called, the default name of the file is *acinfo* ISXSAVE A, where *acinfo* is the accounting information or initials entered at log-on. The CMS command "`LIST * ISXSAVE`" will list the CMS file identifiers of ISXSAVE files that currently exist on the disk. Unneeded files can be erased using the CMS `ERASE` command:

```
. list * isxsave
BRW ISXSAVE A
SMK ISXSAVE A
R;

. erase brw isxsave
```

CHAPTER 4

Elements of the ISX Language

The Interactive STATLIB Executive System—ISX—consists of three classes of statements: ISX commands, ISX procedures, and STATLIB subroutines. An awareness of the distinctions among the various classes of ISX statements is useful in clarifying differences in syntax. An understanding of the distinctions becomes necessary when one consults the two reference manuals that provide companion documentation to this primer.

ISX COMMANDS

The primary function of ISX commands is to control the operation of the ISX system. ISX commands always start with a slash (/). Most ISX commands can be abbreviated. The ISX commands we have discussed so far include the following (with abbreviations in parentheses):

- To end an ISX session:

```
/LOG
/STOP  (/S)
```

- To save and restore dynamic core:

```
/SAVE  (/SAV)
/RESTORE  (/RES)
```

- To create ISX variables:

```
/DEFINE  (/DEF)
/DEFAI
/DEFAR
```

- To print a line or message:

```
/
```

- To list existing DYNCOR block identifiers:

```
/LIST  (/L)
```

Additional ISX commands for creating user-defined procedures, executing ISX commands stored in files, and debugging ISX "programs" are discussed in Chapter 7, Programming in ISX. ISX commands are completely documented in Chapter 4 of the *STATLIB Manual.*

ISX SYSTEM PROCEDURES

ISX *system procedures*, also referred to as commands in this text, are routines written in the ISX language that provide capabilities ranging from data management and display to sophisticated statistical analysis. Most procedures call one or more STATLIB FORTRAN subroutines. Designed to facilitate use in an interactive environment, system procedures are mostly keyword-driven, consisting of a command verb followed by a sequence of optional *keywords.* In general, these keywords (and their related arguments) can be specified in any order.

System procedures are grouped in packages according to their function: general procedures, time series analysis procedures, and two-way matrix analysis procedures.

General Procedures (GENPROCS)

General procedure commands provide generalized capabilities for ISX session control, as well as input and output functions. These procedures are all members of the package GENPROCS. Some examples of general procedure commands we have already discussed include:

TERM	to initialize ISX for interactive processing
INPUT	to retrieve data stored in external files
OUTPUT	to store data in external files

Additional useful general procedure commands are discussed in Chapter 8, Utility Routines. General procedures are completely documented in Chapter 2 of the *Procedure Manual.*

Time Series Analysis Procedures (TSPROCS)

Time series analysis procedures provide a wide range of capabilities for manipulating and analyzing time series data. These commands are all members of the package TSPROCS. Most time series procedure commands end with the suffix TS. Examples of typical time series procedures are:

PRTTS	to print time series data
PLOTTS	to plot time series data
REG	to perform regression analysis on time series data

Many commonly used time series procedures are described in Chapter 5, which is on Time Series Analysis and Forecasting. Time series procedures are completely documented in Chapter 4 of the *Procedure Manual.*

Two-Way Matrix Analysis Procedures (STPROCS)

Two-way matrix analysis procedures provide many capabilities for managing and analyzing two-way matrix data. These commands are members of the package STPROCS. Most STPROCS commands end with the suffix ST. Examples of typical matrix procedures are:

PRTST	to print matrix data
PLOTST	to plot matrix data
REGST	to perform regression analysis on matrix data

Chapter 6, on Two-Way Matrix Analysis, describes many commonly used two-way matrix procedures. These procedures are completely documented in Chapter 3 of the *Procedure Manual.*

User Defined Procedures

ISX also provides the capability for users to develop their own procedures or "macros" by grouping together a set of ISX statements. Once a procedure is defined, its name becomes a valid ISX command. Commands for creating user-defined procedures are described in Chapter 7, on Programming in ISX.

STATLIB SUBROUTINES

STATLIB subroutine calls comprise the third class of statements in ISX. Written in the FORTRAN language, STATLIB subroutines form the core of the STATLIB/ISX system.

While most STATLIB subroutines can be called from within ISX, system procedure calls can be used in place of many to perform the same function (usually by calling the STATLIB subroutine). For the average ISX user, the procedure version, if it exists, will be easier to use and is therefore emphasized in this primer.

STATLIB subroutine calls provide additional functions required by more advanced ISX users that allow greater control over computations and manipulation of data. Subroutine calling sequences are free of punctuation but fixed in the number and order of arguments. In fact, subroutine calls from ISX are virtually identical to the actual subroutine calls used to access the STATLIB library directly from FORTRAN programs.

STATLIB subroutines are also grouped in packages according to function, including time series analysis, two-way matrix analysis, and several others.

Time Series Analysis Subroutines (TSPAK)

STATLIB time series subroutines are provided to manage and model time series data. Members of the package TSPAK, these routines all begin with the prefix TS. Several useful TSPAK subroutines are included in the next chapter, on Time Series Analysis and Forecasting. TSPAK routines are completely documented in Chapter 6 of the *STATLIB Manual.*

Two-Way Matrix Analysis Subroutines (STPAK)

STATLIB two-way matrix subroutines manipulate and analyze two-way matrix data. These routines, members of the package STPAK, all begin with the prefix ST. A few useful STPAK subroutines are included in Chapter 6 of this text, on Two-Way Matrix Analysis. STPAK routines are completely documented in Chapter 5 of the *STATLIB Manual.*

Additional STATLIB Packages

Several other STATLIB subroutine packages are accessible via ISX, and are also documented in Part II of the *STATLIB Manual.*

The SERVICE package includes subroutines to set options and parameters for routines indirectly called via ISX. Some useful SERVICE routine calls are included in Chapter 8, on Utility Routines.

The DYNCOR package contains routines that manage dynamic core storage. The SPPAK package provides subroutines to perform spectrum analysis of time series data. TABLE is a package of routines for managing and analyzing multiway tables of data. These routines are beyond the scope of this primer.

THE `HELP` COMMAND

The `HELP` command displays the calling sequences of specified ISX statements. This convenient on-line capability provides a useful memory aid when working at the terminal. The `HELP` command assumes various forms, depending on the desired function.

`HELP PACKAGES` lists all the STATLIB/ISX package names:

```
> help packages
PACKAGES : GENPROCS STPROCS TSPROCS ISX STPAK TSPAK
  SPPAK TABLE DYNCOR SERVICE
```

`HELP packagename` lists all the commands or routines included in the specified package. For example:

```
> help genprocs
GENPROCS :
  CLSPRFI  DELETE  HELP  INPUT  OO  OPTLIST
  OUTPRFI  OUTPUT  TERM
```

`HELP packagename command...` lists the calling sequence of the specified command(s) or routine(s) within the designated package—For example:

```
> help genprocs input output
INPUT [NOPRINT] filename [FILETYPE ft] [FILEMODE fm]
  [typ1 nam1 typ2 nam2 ..]
INPUT [NOPRINT] NOHEADER filename [FILETYPE ft]
  [FILEMODE fm] typ1 nam1 LEN1 fmt1 ..
OUTPUT [NOPRINT] [REPLACE] [NOHEADER] [FORMAT AUTO/fmt]
  filename [FILETYPE ft] [FILEMODE fm] typ1 nam1 ..
```

`HELP packagename * command...` lists all the commands in the specified package, and the calling sequence of the indicated command(s). For example:

```
> help tsprocs * namets prtts
TSPROCS :
  ACPLT    CLIM   C-O    CORRTS   DEPV    DURBINH
  ENTER    EXSM   FCST   FILE     FUPD    G-QTS
  H-L      LOAD   M-L    NAMETS   PDL     PLOTTS
  PRTTS    REG    REGSP  REPLACE  RWRREG  SERIES
  SET      TPA    TRAN   TSTOST   2S      2SF
  2SF1BY1  2SLS   2SM
NAMETS [T] [typ1 nam1 typ2 nam2 ..]
PRTTS [TIME N1Y N1M N2Y N2M] [INDIC type name]
  [TOTALS] [FORMAT fmtcod] typ1 nam1 typ2 nam2 ..
```

`HELP` with no arguments prints the calling sequence of the `HELP` command.

ISX SYSTEM STRUCTURE

Figure 4.1 illustrates the relationship of the primary packages comprising the ISX system.

Chapter 5 describes many commonly used time series analysis and forecasting commands (and subroutines). Chapter 6 describes two-way matrix analysis commands (and subroutines). These chapters are somewhat parallel in content. Users primarily interested in time series analysis and forecasting may wish to read Chapter 5 first; users analyzing two-way matrix data may begin with Chapter 6.

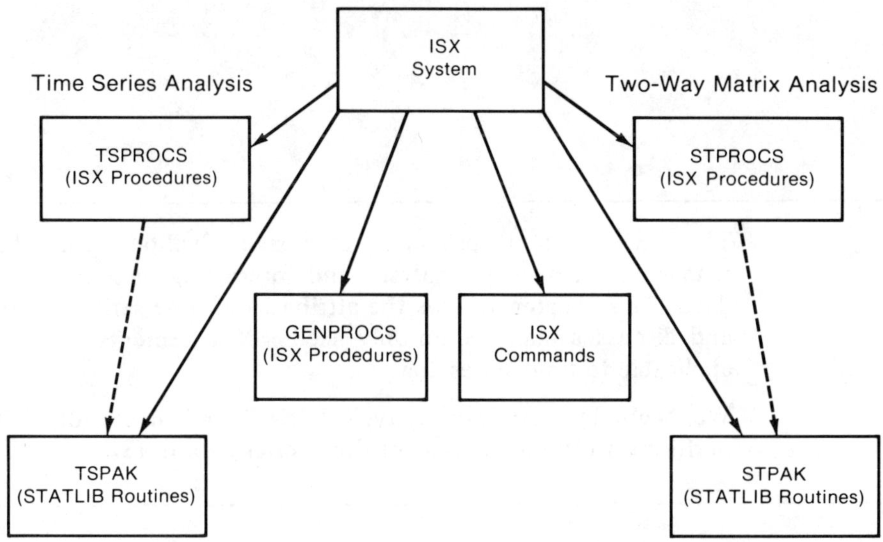

Figure 4.1. The primary packages making up the ISX system.

CHAPTER 5

Time Series Analysis and Forecasting

The ISX system provides extensive facilities for the management, display, analysis, and forecasting of time series data. This chapter reviews the attributes of time series in ISX and discusses many commonly used ISX statements that are applicable to time series analysis.

We begin by examining a typical ISX terminal session that performs a routine analysis of time series data in ISX.

A SAMPLE ISX SESSION

The following example demonstrates the sequence of interactive commands that might be used to begin an analysis of the relationships among monthly housing-starts, revenue, and message unit data for a hypothetical telephone company. (These series, <DATA HOUS>, <DATA REV>, and <DATA MSG>, which are stored in a file called EXAMPLE DATA A, will be referred to throughout this chapter.) Our intent at this point is not to dwell on the details of each statement, or to develop an optimal model. Rather, this example is included to illustrate the interactive, sequential nature of ISX processing. Each statement is entered and executed immediately. A statement might create new data in core that would then be available in subsequent steps; or the results of one statement might be interpreted by the user to determine the next logical step in the analysis. Thus, each ISX statement can be thought of as a building block in the analysis sequence.

```
> term
> input example
INPUT FROM EXAMPLE DATA * ...
  DYNINP: 28 .DAT REV
  DYNINP: 28 DATA REV
  DYNINP: 28 .DAT MSG
  DYNINP: 28 DATA MSG
  DYNINP: 28 .DAT HOUS
  DYNINP: 28 DATA HOUS

> prtts time 1981 1 1981 8 data rev data msg data hous
STPAK / STPRT

 MATRIX=TMP.    OBS    1-    8

OBL           DATAREV   DATAMSG   DATAHOUS

JAN 1981        50.69     15.12      73.90
FEB 1981        61.55     17.05      70.20
MAR 1981        57.78     16.78     125.10
APR 1981        63.42     16.93     142.30
MAY 1981        60.59     16.50     170.50
JUN 1981        59.67     15.54     174.10
JUL 1981        57.71     17.09     144.50
AUG 1981        59.67     16.89     146.80

> set data lrev = log ( data rev )
TSFN1:  DTA. 1 = EQL(DATA REV), JAN 1971-DEC 1981

TSFN1:  DTA. 0 = LOG(DTA. 1), JAN 1971-DEC 1981

TSFN1:  DATA LREV = EQL(DTA. 0), JAN 1971-DEC 1981
```

```
> plotts time 1980 1 1981 8 data lrev ; data hous
TSPAK  / TSPLT
```

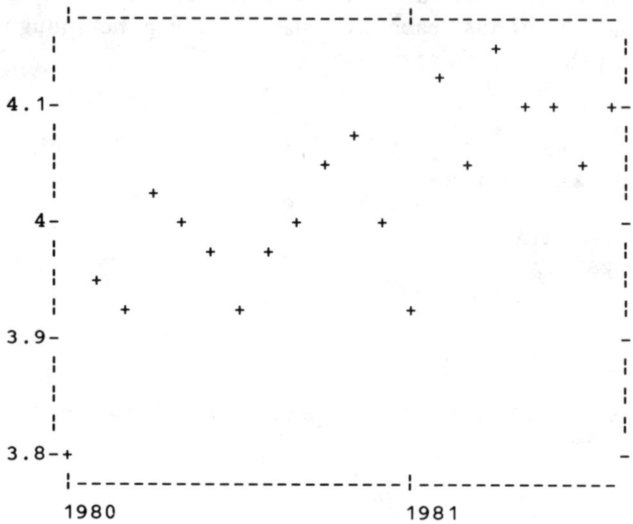


```
1 + DATALREV  VS  TIME
JAN 1980-AUG 1981

TSPAK  / TSPLT
```

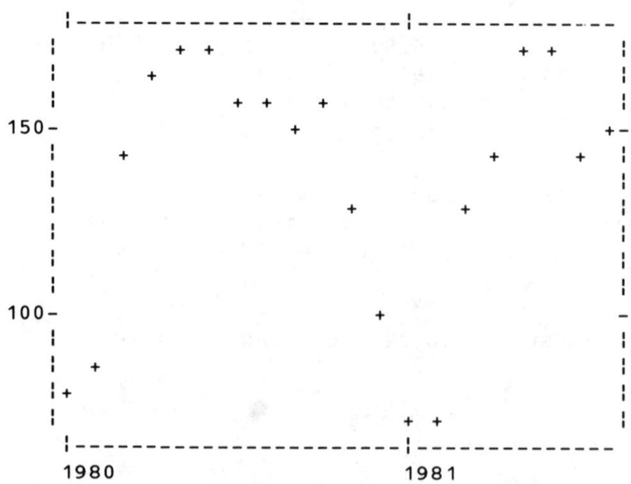


```
1 + DATAHOUS  VS  TIME
JAN 1980-AUG 1981
```

```
> plotts data lrev : data hous
TSPAK  / TSPLT

    ----|---------|---------|---------|---------|-------
   |                                   +                 |
   |+                                                    |
4.1-                                      +           ++ -
   |                            +                        |
   |                           +         +     +         |
   |                                               +     |
  4-                +                      +            +-
   |                                          +        + |
   |       +                                             |
   | +                                  +     +          |
3.9-                                                     -
   |                                                     |
   |                                                     |
   |                                                     |
3.8-  +                                                  -
    ----|---------|---------|---------|---------|-------
         80        100       120       140       160

   1 + DATALREV  VS  2 DATAHOUS
   JAN 1980-AUG 1981
```

TIME SERIES STRUCTURE

As we have seen, a time series in ISX is a data structure containing an ordered sequence of observations over time (or space). ISX can handle series of any periodicity, including daily, weekly, monthly, quarterly, and annual, to name a few.

DYNCOR Blocks

Each series consists of two DYNCOR blocks: a *data block* containing the actual data values, and an associated *time control block* containing the starting date and periodicity corresponding to the data (and possibly a descriptive identification of the series). Each block is identified by a blocktype and a blockname. Both blocks associated with a series are assigned the same one- to four-character blockname. This blockname is often referred to as the *series name* in ISX. The blocktype of the data block also consists of one to four characters, while the blocktype of the corresponding time control block consists of the first three characters of the data blocktype, preceded by a period—For example:

<DATA MSG>	MSG data block
<.DAT MSG>	MSG time control block
<PRDN REV>	REV (predictions) data block
<.PRD REV>	REV (predictions) time control block

Most ISX statements (procedure commands and STATLIB subroutines) require specification of both the series blocktype and blockname. Some require that only the series name be specified and require a specific blocktype.

Time Control Information

Three items contained in the time control block define the starting date and periodicity of the associated time series data. They are the starting *year* ("year" referring to major time periods), the starting *month* ("month" referring to minor time periods), and the number of *months per year*. Annual (or undated) observations can be defined as having one month per year.

CREATING TIME SERIES IN ISX

A time series can be created in the ISX environment in several ways. A series can be entered interactively within ISX, retrieved from an existing external file, or calculated as a function of other series. Once defined in ISX, a series remains in core for the duration of the ISX session unless modified or deleted. A newly created series can be referred to in any subsequent commands during the session.

Entering Series Interactively within ISX

The `ENTER` command can be used to create a new series interactively within ISX. `ENTER` creates both the data block and the time control block for the new series.

The following example creates a new series, <DATA QREV>, which contains quarterly revenue data from the third quarter of 1980 through the second quarter of 1982:

```
> enter data qrev torg 1980 3 4 1200 1400 1300 $
> 1400 1500 1600 1450 1700
```

Here "data qrev" is the series type and name, and "torg 1980 3 4" specifies the starting year (1980), the starting month (3—third quarter), and the number of months per year (4—quarterly). The remaining eight entries are the data values for third quarter 1980 through second quarter 1982. Note the use of the $ argument to continue input on the next line.

The general form of the `ENTER` command is as follows:

```
ENTER blocktype blockname TORG start-year
   start-month months-per-year value1 value2..
```

A second format of the `ENTER` command permits specification of descriptive identification information to be stored in the time control block—For example:

```
> enter data qrev torg 1980 3 4 ident quarterly $
>    revenue data
```

In this case the identification information "quarterly revenue data," preceded by the keyword IDENT, replaced the data values in the `ENTER` statement. Since `ENTER` can accommodate data values or IDENT and its arguments, but not both, a subsequent `ENTER` command must be issued to input the data values—For example:

```
> enter * 1200 1400 1300 1400 1500 1600 1450 1700
```

Here, the * is an abbreviated form indicating that this `ENTER` command refers to the series specified in the preceding `ENTER` command.

`ENTER` is part of the time series data management facility discussed near the end of this chapter.

Reading Series from a File

The `INPUT` command can be used to retrieve time series stored in data files and input them into ISX. A data file containing time series (time control blocks and data blocks) can be created in the EDIT environment or generated from within ISX. (Chapter 3, on Data in ISX, includes format descriptions for creating data files.)

A data file may contain many blocks. `INPUT` can copy the entire contents of a file into ISX:

```
> input example
INPUT FROM EXAMPLE DATA * ...
  DYNINP: 28 .DAT REV
  DYNINP: 28 DATA REV
  DYNINP: 28 .DAT MSG
  DYNINP: 28 DATA MSG
  DYNINP: 28 .DAT HOUS
  DYNINP: 28 DATA HOUS
```

or retrieve only selected blocks from a file:

```
> input example data hous data rev
INPUT FROM EXAMPLE DATA * ...
  DYNINP: 28 .DAT HOUS
  DYNINP: 28 DATA HOUS
  DYNINP: 28 .DAT REV
  DYNINP: 28 DATA REV
```

Verification of blocks retrieved may be suppressed by including the NOPRINT argument in the `INPUT` command:

```
> input noprint example
```

If a series data block is retrieved without a time control block, its time control block can be created via the `TSTORG` routine. For example:

```
> tstorg data msg 1978 1 12
```

creates the time control block, <.DAT MSG>, for a monthly series (12) starting in January 1978 (1978 1).

The `LOAD` command can also be used to retrieve copies of time series blocks saved on a file and input them into ISX. `LOAD` is included in the time series data management facility described near the end of this chapter.

SAVING A TIME SERIES

A series created or modified during an ISX session is no longer available once ISX is exited. The new series must be saved in a file if it is to be accessed during a later session. The `OUTPUT` command can be used to record copies of series within ISX on a file for permanent storage. The contents of the file can then be accessed in subsequent sessions using the `INPUT` command described previously.

In the following example, a new series, <DATA RPM> (revenue per message), is calculated as the quotient of <DATA REV> and <DATA MSG> using the `SET` procedure. The new series is then appended to the file EXAMPLE DATA A so that it will be available for analysis at a later time.

```
> set data rpm = data rev / data msg
TSFN2:   DTA. 0 = DIV(DATA REV,DATA MSG LAG 0)
         JAN 1971-DEC 1981

TSFN1:   DATA RPM = EQL(DTA. 0), JAN 1971-DEC 1981
> output example data rpm
OUTPUT (APPEND) TO EXAMPLE DATA ...
  DYNOUT:   .DAT RPM 28
  DYNOUT:   DATA RPM 28
```

The time control block is not specified in the `OUTPUT` command. It is automatically saved along with the data block. If the file EXAMPLE DATA A does not already exist, it will be created.

By default, the data are written to the file in machine-readable format. The *default format* is recommended for compactness, efficiency, and accuracy, but such files cannot be listed or edited at the operating system level. To create a listable file, the FORMAT argument can be included in the `OUTPUT` command.

FORMAT AUTO automatically determines a reasonable output format for the data:

```
> output format auto example data rev
```

Or a desired FORTRAN format can be specified. For example:

```
> output format r(10f7.0) example data rev
```

will generate 10 data values per line, in F7.0 format. See the *Procedure*

Manual for a complete description of OUTPUT's capabilities.

The FILE command can also be used to store time series in a permanent file. FILE is included in the time series data management facility described near the end of this chapter.

DELETING SERIES FROM THE ISX ENVIRONMENT

Each series created or retrieved in ISX occupies space in dynamic core. Series no longer needed can be deleted, making space available for other uses. The DELETE TS command can be used to delete all or specified series from core. For example:

```
> delete ts
```

deletes all time series, while

```
> delete ts data rpm data abcd
```

deletes series <DATA RPM> and <DATA ABCD>.

The global argument "...." can also be used to delete all series with a specified blockname, regardless of blocktype:

```
> delete ts .... rpm
```

or all series with a specified blocktype, regardless of blockname:

```
> delete ts prdn ....
```

NAMETS, in the following section, is useful for verifying which series will be deleted by a global delete command.

LISTING NAMES AND RANGES OF TIME SERIES

The NAMETS command can be used to list the names of all series that currently exist within the ISX environment—For example:

```
> namets
DATA REV DATA MSG DATA HOUS DATA QREV
```

If the keyword T is included as the first argument, the date ranges (and identification information, if present) are displayed along with the series names—For example:

```
> namets t data ....
DATA REV  JAN 1971 - DEC 1981
DATA MSG  JUL 1971 - DEC 1981
DATA HOUS JAN 1972 - AUG 1981
DATA QREV 1980:3 - 1982:2 QUARTERLY REVENUE DATA
```

UPDATING TIME SERIES

Two commands, **ENTER** and **REPLACE**, are useful for updating time series in the ISX environment.

ENTER can be used to append additional values to the end of an existing series, extending the length of the series by the number of values specified. Assume, for example, that the series <DATA RATE> contains average mortgage rates through December 1981:

```
> namets t data rate
DATA RATE JAN 1970-DEC 1981 MORTGAGE RATES
```

The following **ENTER** command appends new values for January through April 1982:

```
> enter data rate 17.25 17.00 16.75 17.25
> namets t data rate
DATA RATE JAN 1970-APR 1982 MORTGAGE RATES
```

The **REPLACE** command can be used to replace current values in a series with new values (or to insert or delete one or more values at any point in the series).

REPLACE is useful for correcting erroneous values in a series. For example, the following series contains an incorrect data value for March 1981:

```
              DATA EXAM

JAN 1981        100.
FEB 1981        200.
MAR 1981       3999.
APR 1981        400.
MAY 1981        500.
JUN 1981        600.
```

The command:

```
> replace data exam 1981 3 1 300
```

replaces the erroneous value for March 1981 (1981 3) with one (1) correct value (300).

Similarly, **REPLACE** can be used to replace estimated figures with actual values (as they become available). The series <DATA COST> contains budgeted cost figures for the first six months of 1982:

```
                    DATA COST

JAN 1982              5400.
FEB 1982              5600.
MAR 1982              6200.
APR 1982              6500.
MAY 1982              6700.
JUN 1982              7000.
```

The following command replaces the budgeted values by actual cost values for three months beginning in January 1982 (1982 1):

```
> replace data cost 1982 1 3 5297 5860 6425
```

This results in the updated series:

```
                    DATA COST

JAN 1982              5297.
FEB 1982              5860.
MAR 1982              6425.
APR 1982              6500.
MAY 1982              6700.
JUN 1982              7000.
```

REPLACE can also be used to add values at the beginning of a series, as illustrated by the following example:

```
> namets t data sale
DATA SALE JAN 1971-DEC 1981 WIDGET SALES
> replace data sale 1970 11 2 7500 8600
> namets t data sale
DATA SALE NOV 1970-DEC 1981 WIDGET SALES
```

In this case REPLACE is used to add two new values (7500, 8600) at the beginning of the series, moving the origin from January 1971 to November 1970.

Any modifications made to a series by use of ENTER or REPLACE exist only within the current ISX environment. The modified series must be saved in a permanent file if it is to be accessed during a later ISX session.

Both ENTER and REPLACE are part of the time series data management facility described near the end of this chapter.

PRINTING TIME SERIES DATA

The PRTTS command can be used to print time series data for one or more series over a specified time period.

If a single series is specified, the output is usually displayed in table format. In the following example, the entire series <DATA REV> is printed from January 1980 to December 1981.

```
> prtts time 1980 1 1981 12 data rev
TSPAK / TSPRT            DATA REV

                1980      1981

          1  45.2381  50.6873
          2  52.5833  61.5475
          3  50.8695  57.7839
          4  56.0783  63.4201
          5  54.7997  60.5892
          6  53.1094  59.6725
          7  50.9745  57.7075
          8  53.2653  59.6725
          9  54.2137  59.9738
         10  56.8922  63.4583
         11  58.6955  65.1007
         12  54.1424  60.5001
        SUM 640.8619 720.1133
```

Multiple series are displayed in column format with date labels. The following example prints data values for both <DATA MSG> and <DATA REV> for the year 1981:

```
> prtts time 1981 1 1981 12 data msg data rev
STPAK / STPRT

 MATRIX=TMP.   OBS    1-   12

OBL        DATAMSG   DATAREV

JAN 1981     15.12     50.69
FEB 1981     17.05     61.55
MAR 1981     16.78     57.78
APR 1981     16.93     63.42
MAY 1981     16.50     60.59
JUN 1981     15.54     59.67
JUL 1981     17.09     57.71
AUG 1981     16.89     59.67
SEP 1981     18.21     59.97
OCT 1981     17.75     63.46
NOV 1981     18.18     65.10
DEC 1981     18.10     60.50
```

The preceding example can also include a user-specified format, such as the following:

```
> prtts time 1981 1 1981 12 format f9.0 data msg $
>    data rev
STPAK / STPRT

 MATRIX=TMP.    OBS    1-   12

OBL         DATAMSG   DATAREV

JAN 1981         15.        51.
FEB 1981         17.        62.
MAR 1981         17.        58.
APR 1981         17.        63.
MAY 1981         16.        61.
JUN 1981         16.        60.
JUL 1981         17.        58.
AUG 1981         17.        60.
SEP 1981         18.        60.
OCT 1981         18.        63.
NOV 1981         18.        65.
DEC 1981         18.        61.
```

When the TOTALS keyword is included, PRTTS prints yearly totals along with the data, if it is printing more than one series—For example:

```
> prtts time 1980 1 1981 12 totals data msg data rev
```

The PRTTS command requires specification of the *time interval* for which the command will apply. In the previous examples, the time period has been given explicitly through inclusion of the TIME argument:

```
TIME start-year start-month end-year end-month
```

The time interval can also be given implicitly, as explained in the next section.

DEFINING TIME INTERVALS FOR TIME SERIES COMMANDS

The PRTTS command described in the previous section is one of many time series procedure commands (and STATLIB subroutines) requiring specification of a time interval for which the command is to apply. Procedure commands containing explicit specification of time intervals adhere to the following general format:

```
cmd TIME start-year start-month end-year end-month ...
```

while STATLIB subroutine calls containing time interval specifications can be generalized as follows:

```
routine ... start-year start-month end-year end-month
```

A special ISX variable called $T can be defined to contain the time interval that is to apply to subsequent procedure commands. $T can be defined via the ISX `/DEFINE` command. For example:

```
> /define $t 1980 1 1981 12
```

specifies the interval 1980 month 1 (January) through 1981 month 12 (December).

$T is also automatically defined—or redefined—when the TIME argument is included in a command. For example, the command:

```
> prtts time 1980 1 1981 12 data msg
```

automatically defines $T by internally issuing the command:

```
/DEFINE $T 1980 1 1981 12
```

The contents of $T can be verified by issuing the / (slash) command:

```
> / $t
1980 1 1981 12
```

Once $T is defined, the TIME argument can be eliminated in subsequent procedure commands requiring time interval specification. The interval will be defined by the current contents of the variable $T. Thus, if $T contains "1980 1 1981 12," the procedure command:

```
> prtts time 1980 1 1981 12 data rev
```

can be shortened to:

```
> prtts data rev
```

STATLIB subroutine calls can be shortened by including the variable $T in place of the actual time interval specification. ISX will substitute the current contents of $T when processing the statement. Thus, the

STATLIB routine to plot a series versus time:

```
> tsplt 1 data rev 1980 1 1981 12
```

can be shortened to:

```
> tsplt 1 data rev $t
```

(The STATLIB routine `TSPLT` has been included to illustrate the use of $T. Most users will probably use the procedure command `PLOTTS`, discussed in this chapter's section on plotting time series data, for generating plots of time series data.)

TIME SERIES TRANSFORMATIONS

The analysis process often requires the creation of new series as functions of existing series and constants. ISX provides extensive capabilities for performing arithmetic transformations, constant and moving functions, and extraction and concatenation on time series data within the ISX environment.

This section describes the most commonly used commands and STATLIB routines for performing transformations and manipulating time series. When new series are created, the corresponding time control blocks are also created. The resulting new series exist only within the current ISX environment and must be stored in a permanent file if they are to be accessed during a later session.

Arithmetic Transformation

The `SET` command performs multiple transformations with one or more existing series (and/or constants) and stores the result in a new time series. The new series is calculated as the result of an observation-by-observation evaluation of an algebraic expression.

The following simple example illustrates the use of `SET` to create a new series, <DATA TOT> (total revenues), equal to the sum of the two series, <DATA BUS> (business revenues) and <DATA RES> (residential revenues).

```
> set data tot = data bus + data res
TSFN2:   DTA. 0 = ADD(DATA BUS,DATA RES LAG 0)
         JAN 1982-JUN 1982

TSFN1:   DATA TOT = EQL(DTA. 0), JAN 1982-JUN 1982
```

Printing the three series displays the result:

```
> prtts time 1982 1 1982 6 data bus data res data tot
STPAK / STPRT

 MATRIX=TMP.    OBS    1-    6

OBL         DATABUS   DATARES   DATATOT

JAN 1982     100.00     50.00    150.00
FEB 1982     200.00    100.00    300.00
MAR 1982     300.00    150.00    450.00
APR 1982     400.00    200.00    600.00
MAY 1982     500.00    250.00    750.00
JUN 1982     600.00    300.00    900.00
```

SET calculates the resulting series for the largest time period common to all series in the expression. The time frame of the series <DATA TOT>, just created, is identical to the two series used to create it.

```
> namets t data bus data res data tot
  DATA BUS   JAN 1982-JUN 1982
  DATA RES   JAN 1982-JUN 1982
  DATA TOT   JAN 1982-JUN 1982
```

In the following example, a new series, <DATA PROF> (profit), is calculated as a function of existing series <DATA INC> (income) and <DATA EXPN> (expenses):

```
> set data prof = data inc - data expn
TSFN2:   DTA. 0 = SUB(DATA INC,DATA EXPN LAG 0)
         JAN 1982-APR 1982

TSFN1:   DATA PROF = EQL(DTA. 0), JAN 1982-APR 1982
```

Printing the entire contents of the three series displays the result:

```
> prtts time 1981 11 1982 6 data inc data expn data prof
STPAK / STPRT

 MATRIX=TMP.   OBS    1-    8

OBL         DATAINC  DATAEXPN DATAPROF

NOV 1981    1700.00       0.0       0.0
DEC 1981    1800.00       0.0       0.0
JAN 1982    1900.00   1100.00    800.00
FEB 1982    2000.00    900.00   1100.00
MAR 1982    2100.00   1200.00    900.00
APR 1982    2200.00   1400.00    800.00
MAY 1982        0.0   1600.00       0.0
JUN 1982        0.0   1500.00       0.0
```

The time frame for the resulting new series, <DATA PROF>, is the time frame common to the series used to create it.

```
> namets t data inc data expn data prof
  DATA INC   NOV 1981-APR 1982
  DATA EXPN  JAN 1982-JUN 1982
  DATA PROF  JAN 1982-APR 1982
```

The general form of the `SET` command is:

```
SET type name = expression
```

where "type name" identifies the new series to be created. The expression may contain a combination of series identifiers; constants (or ISX variables containing constants); the arithmetic operations + (addition), − (subtraction), * (multiplication), / (division), and ** (raise to power), and the functions SQR (square root), LOG (natural logarithm), and EXP (exponentiation base *e*). Evaluation of the expression proceeds from left to right, unless parentheses force a different order. All arguments, including parentheses and operator symbols, must be separated by spaces.

The following examples provide additional illustrations of the `SET` command.

The average of two series can be calculated:

```
> set data ave = ( data rev1 + data rev2 ) / 2
```

The percent of total can be calculated:

```
> set data pctx = ( data x / $tot ) * 100.
```

Here $TOT is a variable containing the series total; it can be created using routine `TSFNTS`, described later in this section.

`SET` is a procedure command that invokes the STATLIB routines `TSFN0`, `TSFN1`, and `TSFN2`. These routines can also be executed directly from ISX, providing access to an extensive list of additional useful functions.

The `TRAN` command can also be used to create a new series as a result of the arithmetic transformation of an existing series. `TRAN` provides the additional feature of "remembering" the transformation that occurred. When a series created via `TRAN` is used as a dependent variable in a subsequent regression or forecasting command (such as `REG` or `FCST`, discussed in the regression analysis and forecasting sections later in this chapter), the resulting forecasts and residuals are automatically transformed back to the scale of the original series. (For transformations not requiring this feature, `SET` should be used in place of `TRAN`.)

`TRAN` provides four transformation functions particularly useful for developing regression and forecasting models. They are LOG (natural logarithm), SQR (square root), DIFF *k* (*k*th order of differencing), and BXCX (Box-Cox transformation).

The following example illustrates the use of the `TRAN` command:

```
> tran hous lhou log
TSFN1:   DATA LHOU = LOG(DATA HOUS), JAN 1972-AUG 1981
```

In this example, `TRAN` creates a new series, <DATA LHOU>, containing the natural logarithm of <DATA HOUS>. The command

```
> tran rev drev diff 4
TSDIFF: DATA DREV = 4 DIFF OF DATA REV, MAY 1971-DEC 1981
```

creates a new series <DATA DREV>, containing the fourth-order differences of <DATA REV>.

Moving Functions

STATLIB subroutine `TSFNM` can be used to create a new series as a *moving function* (average, sum, etc.) of an existing series. Each element of the new series is a function of *n* points in the original series.

The available functions are as follows:

Function	Result
AVG	Average
EXT	Extract *C*th element
MAX	Maximum
MED	Median
MIN	Minimum
ORD	*C*th Order statistic
SSQ	Sum of squares
STD	Standard deviation
SUM	Sum
VAR	Variance

The desired function can be evaluated for each set of *n* adjacent points in the original series, creating a new series of the same periodicity as the original series. Or the function can be evaluated once for each *n*-point non-overlapping segment, creating a new series of shorter periodicity than the original series (e.g., creating a quarterly or annual series from monthly data).

A position code (L—left, R—right, C—center, S—segment) specifies the data placement of each new series element relative to each original *n*-point segment.

Left, right, and center create new series of the same periodicity as the original series. The date range of the resulting series depends upon both the position code and the specified segment length. Functions are not evaluated for partial segments.

R (right) specifies that each resulting element correspond to the date of the last element of each *n*-point segment in the original series. For example, given the original series <DATA REV>:

```
> namets t data rev
  DATA REV  JAN 1971-DEC 1981
```

the command:

```
> tsfnm data revs sum data rev 12 r
TSFNM:  DATA REVS = 12-PT R SUM(DATA REV), DEC 1971-DEC 1981
```

creates a new monthly series, <DATA REVS>, starting in December 1971, which contains right twelve-month moving sums of <DATA REV>.

L (left) specifies that the resulting elements be dated to correspond to the date of the first element of the *n*-point segments. For example:

```
> tsfnm data reva avg data rev 12 l
TSFNM:  DATA REVA = 12-PT L AVG(DATA REV), JAN 1971-JAN 1981
```

creates a new monthly series, <DATA REVA>, starting in January 1971, which contains left twelve-month moving averages of <DATA REV>.

In the following example, resulting maximums for each five-month segment are dated to correspond with the date of the center element of each original segment:

```
> tsfnm data revm max data rev 5 c
TSFNM:  DATA REVM = 5-PT C MAX(DATA REV), MAR 1971-OCT 1981
```

S (segment) specifies that the desired function is to be evaluated once per *n*-point non-overlapping segment, resulting in a new series of shorter periodicity than the original series. For example, the commands:

```
> tsfnm data qrev avg data rev 3 s
TSFNM:  DATA QREV = AVG(DATA REV), JAN 1971-DEC 1981
        SEG LGTH = 3, NEW TIME = 1971:1-1981:4 <4 M/Y>
> tsfnm data yrev sum data rev 12 s
TSFNM:  DATA YREV = SUM(DATA REV), JAN 1971-DEC 1981
        SEG LGTH = 12, NEW TIME = 1971-1981
```

create a new quarterly series, <DATA QREV>, which contains averages for each quarter (three-month segment), and a new annual series, <DATA YREV>, containing sums for each year (twelve-month segment).

With the EXT and ORD functions, the number of the observation to be extracted and the fractional order statistic (between 0 and 1) should be given, respectively, as the last argument. For example:

```
> tsfnm data rv25 ord data rev 12 s .25
TSFNM:  DATA RV25 = ORD(DATA REV), JAN 1971-DEC 1981
        SEG LGTH = 12, NEW TIME = 1971-1981
```

converts <DATA REV> to the annual series <DATA RV25>, which contains the twenty-fifth fractional order statistic of each year's revenue data.

Functions over Time Series

STATLIB subroutine **TSFNTS** can be used to calculate a statistic or evaluate a function over a specified interval of a time series. The available functions, which are the same as those provided by **TSFNM**, are listed in the previous section. The value of the resulting constant is printed in a one-line message and stored (temporarily) in ISX variable $XXX.

In the following example, the command:

```
> tsfnts avg data rev 1977 1 1981 12
TSFNTS: AVG(DATA REV) = 49.73956, JAN 1977-DEC 1981
```

calculates the average monthly revenue over the years 1977 through 1981.

If the result, which is stored in variable $XXX, is to be used in later calculations, it should be saved in another ISX variable, since ISX might subsequently redefine $XXX. The following commands calculate the sum of the elements in series <DATA REV> and save the result in variable $TOT for use later in the session:

```
> tsfnts sum data rev 1971 1 1981 12
TSFNTS: SUM(DATA REV) = 5372.35, JAN 1971-DEC 1981
> /def $tot $xxx
```

Cumulative Sums

STATLIB routine **TSCSUM** creates a new series containing the cumulative sums of an existing series over a specified time interval.

In the following example, **TSCSUM** creates a new series, <DATA CREV>, containing cumulative revenues from <DATA REV> for the period January 1976 through December 1981:

```
> tscsum data rev data crev 1976 1 1981 12
TSCSUM: DATA CREV = CUM SUM OF DATA REV, JAN 1976-DEC 1981
```

Other Time Series Manipulations

It is often useful to create a new time series containing a portion of an existing series, or a concatenation of two contiguous series.

STATLIB subroutine **TSEXT** can be used to extract a portion of an existing series and store it as a new series. The original series remains unchanged.

The following example creates a new series, <DATA EREV>, containing the portion of <DATA REV> from January 1979 through September 1979:

```
> namets t data rev
  DATA REV  JAN 1971-DEC 1981
> tsext data rev 1979 1 1979 9 data erev
TSEXT:  DATA EREV = DATA REV, JAN 1979-SEP 1979
```

Subroutines in DYNCOR, STATLIB's core management system, can also be used to manipulate time series. Care must be exercised, however, since the user is then responsible for checking and maintaining time control blocks.

Additional Routines

STATLIB subroutine **TSPOLY** generates time series containing powers of t (t = 1, 2, 3 ...). **TSHARM** creates series containing sines and cosines. These commands are beyond the scope of this primer.

PLOTTING TIME SERIES DATA

ISX provides extensive facilities for the graphical display of time series data. The **PLOTTS** command generates both *time plots* and *scatter plots*, and the **ACPLT** command produces *autocorrelation plots*. By default, plots are printed at the user's terminal. The GRPAK graphics package, described in Chapter 9, provides additional graphics capabilities. By resetting several ISX parameters, graphical output can be directed to various screen and pen plotting devices.

Time and Scatter Plots

A *time plot* is a plot of a time series on the vertical axis versus time on the horizontal axis. The following example is a time plot of <DATA REV> for the interval January 1979 through December 1981.

```
> plotts time 1979 1 1981 12 data rev
TSPAK   / TSPLT

   |-----------|-----------|-----------
  |                                  + |
  |                           +     +  |
  |                         +          |
60-                            ++ ++  +-
  |                      +   +   +     |
  |               +     +              |
  |                +   +               |
  |             +   + +   +            |
  |          +   +   +                 |
50- + +     +              +           -
  |    ++ ++  +                        |
  |  +                                 |
  |      +     +                       |
  |                                    |
  |+                                   |
   |-----------|-----------|-----------
   1979        1980        1981

  1 + DATAREV   VS   TIME
  JAN 1979-DEC 1981
```

A *scatter plot* is a plot of one or more time series versus another series. To generate a scatter plot, add to the time plot calling sequence a colon (as a delimiter) followed by the identifier of the series to be used as the horizontal variable.

The following example plots a single vertical series, <DATA REV>, against a horizontal series, <DATA MSG>, for the interval January 1980 through December 1981.

```
> plotts time 1980 1 1981 12 data rev : data msg
TSPAK  / TSPLT
```

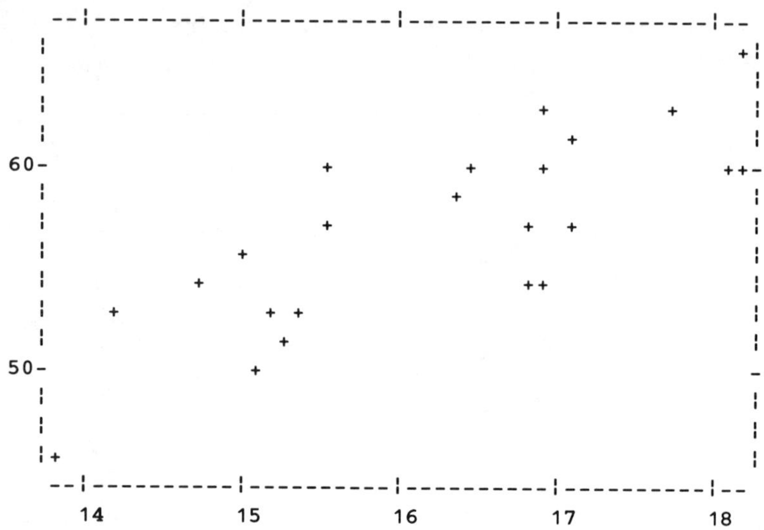

```
1 + DATAREV   VS  2 DATAMSG
JAN 1980-DEC 1981
```

More than one series may be plotted on the vertical axis of a time or scatter plot. In the call to PLOTTS, the series are listed together, separated by ampersands (&). In the following example, both <DATA REV> and <DATA MSG> are plotted (on the vertical axis) versus time (on the horizontal axis) for the indicated time interval:

```
> plotts time 1979 1 1981 12 data rev & data msg
TSPAK   / TSPLT

        |-----------|-----------|-----------
       |                         +      ++ |
    60-                      +  +  ++ ++  +-
       |               ++   ++ +  +   +    |
       |  +      ++  ++  +++     +         |
       | ++ ++ ++  +                       |
       |      +     +                      |
    40-+                                   -
       |                                   |
       |                                   |
       |                                   |
       |                                   |
    20-                               X XX-
       |        XXXX XXXXXXXXXXXXXXXXXXX X  |
       |XXXXXXXX    X                      |
        |-----------|-----------|-----------
        1979        1980        1981

       1 + DATAREV   2 X DATAMSG   VS   TIME
       JAN 1979-DEC 1981
```

In all forms of the PLOTTS command, a series may be lagged by specifying an integer lag following the series identifier. In the following example, <DATA REV> is plotted against <DATA HOUS>, which is lagged two time periods.

```
> plotts data rev : data hous 2
TSPAK  / TSPLT

    -----------|------------------|------------------|
   |+                            +                    |
   |            +                                     |
 60-                     +     ++          +          -
   | +                            +       +           |
   |     +                            +               |
   |                           +    +                 |
   |                      +          + +   +          |
   |  +                                    + +        |
 50-              + +    +                         +  -
   |                                       +  +  ++ + |
   |    +                                             |
   |                                +                +|
   |                                                  |
   |                     +                            |
    -----------|------------------|------------------|
               100                150

      1 + DATAREV   VS   2 -2  HOUS
      JAN 1979-OCT 1981
```

Several plots can be generated by a single PLOTTS command. Vertical and/or horizontal axis series are separated by semicolons; PLOTTS generates a separate plot for each vertical—horizontal pair. The following command generates two separate plots, <DATA REV> versus time, and <DATA MSG> versus time, for the specified time interval.

```
> plotts time 1979 1 1981 12 data rev ; data msg
```

The following command generates four plots:

```
> plotts data brev & data rrev ; data rev : $
>   data msg ; data hous
```

The resulting plots would be as follows:

1. BREV and RREV vs MSG
2. BREV and RREV vs HOUS
3. REV vs MSG
4. REV vs HOUS

Autocorrelation Plots

The **ACPLT** command can be used to plot the *autocorrelation function* and/or *partial autocorrelation function* of one or more time series over a specified time interval.

The following example generates the autocorrelation plot of <DATA REV>:

```
> acplt time 1976 1 1981 12 data rev
TSACOR: DATA REV, 24 LAGS, JAN 1976-DEC 1981

TSPAK  / TSPLT

    ¦---------¦---------¦---------¦---------¦--------
  1-+                                                -
   ¦+ + +                                            ¦
   ¦+ + + + + + +                                    ¦
   ¦+ + + + + + + + +       +                        ¦
   ¦+ + + + + + + + + + + + + +                      ¦
   ¦============================+=+=+=+=+============¦
   ¦+ + + + + + + + + + + + + + + + + + + + + +     +¦
  0---------------------------------------------+-+---
   ¦                                                 ¦
   ¦=================================================¦
   ¦                                                 ¦
   ¦                                                 ¦
   ¦                                                 ¦
   ¦                                                 ¦
 -1-                                                 -
    ¦---------¦---------¦---------¦---------¦--------
    0         5        10        15        20

   1 + ACORREV   VS   TIME
   0-24
```

Approximate 95% confidence limits are plotted as equal signs (=).

Partial autocorrelation plots are obtained by including the keyword PACF before the series names, both autocorrelation and partial autocorrelation plots by including BOTH.

CORRELATION COEFFICIENTS

The CORRTS command can be used to print the *correlation matrix* for a group of time series over a specified time interval.

Only series names are specified (blocktype must be DATA). Lagged series may be indicated by including an integer lag following the series name.

The following example calculates and prints the correlation matrix of series <DATA REV>, <DATA MSG>, and <DATA REV> lagged one period, for the interval January 1976 through December 1980.

```
> corrts time 1976 1 1980 12 rev ; msg ; rev 1
REGPAK / REGCOR


  CORRELATION MATRIX:

                    1         2         3

  1 REV ....    1.0000

  2 MSG ....    0.8593    1.0000

  3 REV ...1    0.7984    0.8054    1.0000
```

If the optional argument DSP is included, CORRTS generates a symbolic *triangular display* in addition to the correlation matrix, as illustrated in the following example:

```
> corrts time 1977 1 1980 2 dsp rev ; msg ; rev 1
REGPAK / REGCOR


   CORRELATION MATRIX:

                     1          2          3

   1 REV ....     1.0000

   2 MSG ....     0.6926     1.0000

   3 REV ...1     0.4602     0.5146     1.0000

TRIPAK / TRIDIS

    1 REV .... X
    2 MSG .... * X
    3 REV ...1 * * X

               1 2 3

   SYMBOLS: -(-.1,-.4) =(-.4,-.7) #(-.7,-1.) <(<-1.)
            +(+.1,+.4) *(+.4,+.7) X(+.7,+1.) >(>+1.)
```

REGRESSION ANALYSIS

The REG command fits various types of regression models (including ordinary least squares, autoregression, moving-average, autoregressive moving-average, and multivariate autoregression) to time series data.

The most basic form of the REG command performs an *ordinary least squares* (OLS) regression of a dependent variable on one or more independent variables. Only the series names of the variables are specified; the blocktype must be DATA. The following example regresses a dependent series, <DATA REV>, on two independent series, <DATA MSG> and <DATA HOUS>, for the specified time interval.

```
> reg time 1976 1 1980 12 rev ; msg ; hous
TSREG:   MODEL=L.S., TYPE=DATA, JAN 1976-DEC 1980
         DEPV=REV

TSPRDN:  MODEL=L.S., TYPE=DATA/PRDN/RSID, JAN 1976-DEC 1980
         DEPV=REV

REG - REGR ANALYSIS..$RGA, RSID PLOT..$RP [IND [LAG]]
- RSID VS PRDN PLOT..$RVP, ACOR PLOT..$AP, PACF PLOT..$PAP
- PRTL RSID PLOT..$PRP IND [LAG]
- PRTL ADJ RSID PLOT..$PAR IND [LAG], REGR COEF CORR..$RCC
- RSID ANALYSIS..$RSA, CHI-SQUARED..$CSQ
- DURBIN-WATSON..$DW, RSID STD NORMAL PLOT..$NRM
```

The optional-output list just given (`$RGA`, `$RP`, etc.) indicates various regression results, statistics, and additional output that can be obtained after executing the `REG` command. Specify the appropriate command (`$RGA`, `$RP`, etc.) to request desired optional output. To generate the regression analysis summary, specify `$RGA`:

```
> $rga
REGPAK / REGANL


  SAMPLE SIZE . . . . . . . . .  60
  SUM OF WEIGHTS . . . . . . . . 6.0000D+01
  ESTIMATED STD DEV . . . . .  2.2471D+00
  R SQUARED . . . . . . . . .  0.8282


  VARIABLE      COEFFICIENT     ESTD STD DEV        T

   0            -1.6130D-01      2.9002D+00     -0.0556
   2 MSG ....    3.0413D+00      2.2594D-01     13.4604
   3 HOUS....    4.0866D-02      7.9857D-03      5.1174

   1 REV ....   DEPENDENT VARIABLE


  ANALYSIS OF VARIANCE

     SOURCE        DF        SS             MS            F

     REGRESSION     2     1.3877D+03     6.9384D+02     137.403
     ERROR         57     2.8783D+02     5.0496D+00

     TOTAL         59     1.6755D+03

 TSREGA: MODEL=L.S., TYPE=DATA, DEPV=REV, JAN 1976-DEC 1980
```

To generate a plot of residuals versus predictions, specify **$RVP**:

```
> $rvp
TSPAK  / TSPLT
```

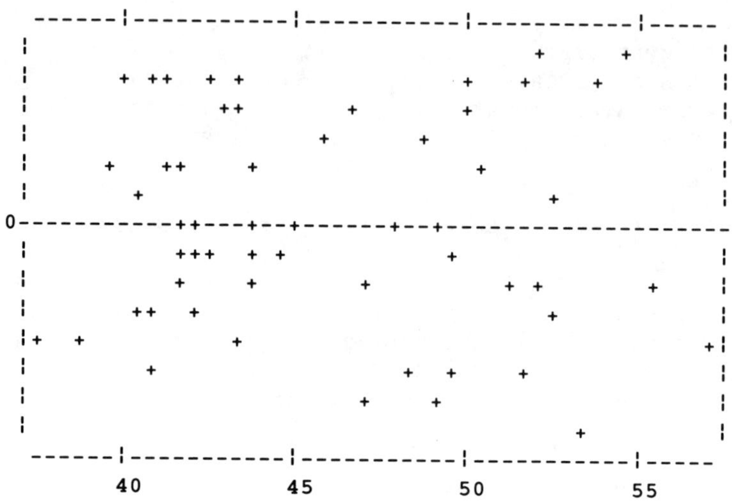


```
1 + RSIDREV   VS  2 PRDNREV
JAN 1976-DEC 1980
```

The following table lists each command and the resulting output:

COMMAND	OUTPUT
`$RGA`	Regression analysis
`$RP [ind (lag)]`	Residuals vs independent variable plot with optional lags (with no independent variable specified, the residuals will be plotted against time).
`$RVP`	Residuals vs predictions plot
`$AP`	Sample autocorrelation plot (ordinary correlogram)
`$PAP`	Sample partial autocorrelation plot (partial correlogram)
`$PRP [ind (lag)]`	Partial residual plot of independent variable with optional lags
`$PAR [ind (lag)]`	Partial adjusted residual plot of independent variable with optional lags
`$RCC`	Regression coefficients correlations
`$RSA`	Residual analysis
`$CSQ`	Chi-squared statistic
`$DW`	Durbin-Watson statistic
`$NRM`	Standard normal plot of residuals

Defining ISX variables with the same names as these commands should be avoided.

`REG` also creates two new series, <PRDN REV> and <RSID REV>, containing predicted (fitted) values and residuals, respectively. Other information pertaining to the model is also stored for later use. (See the following section on forecasting.)

Variables may also be lagged by including one or more lags following a variable name. The example:

```
> reg rev ; msg 1 ; hous 6
```

fits the dependent variable, <DATA REV>, to <DATA MSG> lagged one period and <DATA HOUS> lagged six periods (over the time period defined previously). The example:

```
> reg rev 1 2 ; msg
```

fits <DATA REV> to <DATA REV> lagged one period, <DATA REV> lagged two periods, and <DATA MSG>.

An *autoregressive* (AR) model is a model in which the only independent variables are lagged values of the dependent variable. The following example fits <DATA REV> against <DATA REV> lagged one period and <DATA REV> lagged two periods:

```
> reg rev 1 2
```

A *moving-average* (MA) model may be specified using the special code "M.A." in place of the independent variable, for example:

```
> reg rev ; m.a. 1 2
```

The *autoregressive moving-average* (ARMA) model combines autoregressive and moving-average formats. For example:

```
> reg rev 1 2 ; m.a. 1
```

Additional `REG` command options are available to rename model results (MODL), generate a correlation matrix (CORR), adjust the dependent series for outliers (ROBUST), adjust for autocorrelation of the residuals (REGSTAR), and perform weighted least squares (WGTS). A discussion of these options is beyond the scope of this primer; see the *Procedure Manual* for a detailed description.

FORECASTING

ISX provides commands to generate forecasts and confidence limits based on models generated by the `REG` command described in the previous section.

Executing the `REG` command with dependent variable <DATA dep> creates a new series, <PRDN dep>, which contains predicted values of the dependent variable for the time interval specified in the `REG` command.

The `FCST` command uses the previously developed regression model to generate forecasts (i.e., extend <PRDN dep>) from the time period following the end of the `REG` time interval through the time period specified in the `FCST` command. This is illustrated by Figure 5.1.

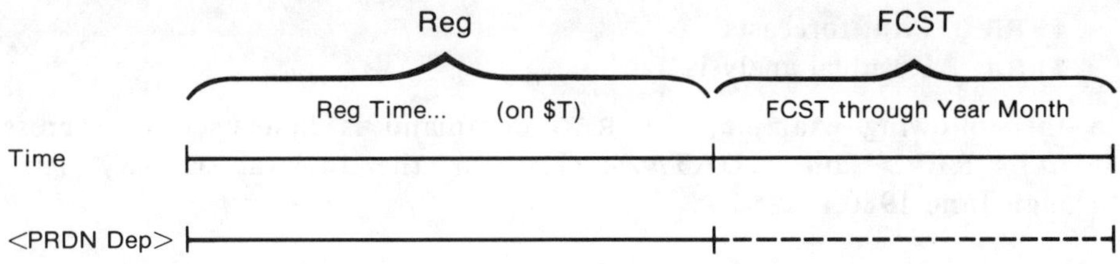

`Figure 5.1.` `FCST` generates forecasts, usually over a time period following the regression period.

The calling sequence of the `FCST` command is simply:

```
FCST THRU year month
```

By default, the model used is the most recent model computed by the `REG` command. (Model names may be defined in the `REG` command and then specified as additional arguments to the `FCST` command.)

Predictions for independent variables required by the model must exist for the forecast period. If a predicted independent variable (<PRDN ind>) does not exist, `FCST` will try to create it by copying <DATA ind>. In this case, independent variables can be extended only to the end of the corresponding data series, limiting the range of the `FCST` command accordingly. (Of course, this restriction does not apply to forecasts from autoregressive models.)

`FCST` also creates (or extends) the series of residuals (forecast errors), <RSID dep>, for that portion of the forecast period over which <DATA dep> exists.

The following optional output can be generated subsequent to the `FCST` command:

$FPR Print forecasts
$RSA Residual analysis

In the following example, the REG command is first used to regress <DATA REV> on <DATA MSG> for the interval January 1971 through June 1980.

```
> namets t data rev data msg
  DATA REV  JAN 1971-DEC 1981
  DATA MSG  JUL 1971-DEC 1981
> reg time 1971 7 1980 6 rev ; msg
TSREG:   MODEL=L.S., TYPE=DATA, JUL 1971-JUN 1980
         DEPV=REV

TSPRDN:  MODEL=L.S., TYPE=DATA/PRDN/RSID, JUL 1971-JUN 1980
         DEPV=REV

REG - REGR ANALYSIS..$RGA, RSID PLOT..$RP [IND [LAG]]
- RSID VS PRDN PLOT..$RVP, ACOR PLOT..$AP, PACF PLOT..$PAP
- PRTL RSID PLOT..$PRP IND [LAG]
- PRTL ADJ RSID PLOT..$PAR IND [LAG], REGR COEF CORR..$RCC
- RSID ANALYSIS..$RSA, CHI-SQUARED..$CSQ
- DURBIN-WATSON..$DW, RSID STD NORMAL PLOT..$NRM
```

Then FCST generates revenue forecasts from July 1980 through December 1981:

```
> fcst thru 1981 12
FCST/2SF: WARNING.. PRDN MSG NOT FOUND; DATA SUBSTITUTED **

TSFN1:   PRDN MSG = EQL(DATA MSG), JUL 1971-DEC 1981

TSPRDN:  MODEL=L.S., TYPE=DATA/PRDN/RSID, JUL 1980-DEC 1981
         DEPV=REV

FCST - RSID ANALYSIS..$RSA, PRINT FCSTS..$FPR ( FOR REV )
```

Since <PRDN MSG> did not exist for July 1980 to August 1981, FCST created it by copying <DATA MSG>. If the specified forecast period extended beyond August 1981, then <PRDN MSG> would have to have been created previously (possibly using another model), and the residuals could only be calculated for the portion of the forecast period for which <DATA REV> existed.

The forecasts can be printed by specifying $FPR:

```
> $fpr
TSPAK / TSPRT            PRDN REV

               1980       1981


           1              51.8050
           2              58.6735
           3              57.7011
           4              58.2339
           5              56.7094
           6              53.3074
           7   52.4568    58.8185
           8   52.0862    58.1147
           9   57.7154    62.7814
          10   53.3520    61.1361
          11   56.2327    62.6629
          12   58.2748    62.3918
         SUM 330.1178   702.3357
```

A so-called residual analysis, a point-by-point comparison of data and forecasts, can be printed by specifying `$RSA`.

Following `FCST`, the `CLIM` command can be used to generate confidence limits. For example, the command:

```
> clim prdn rfsd
TSRFSD: MODEL=L.S., JUL 1980-DEC 1981
        DEPV=REV

TSCLIM: REV, UCL/LCL = PRDN +/- 2 * RFSD

CLIM - CLIM PLOT..$CLP, CLIM & DATA PLOT..$CDP ( FOR REV )
```

creates upper and lower confidence limits (<UCL REV> and <LCL REV>) for observed values of <DATA REV> over the forecast period. The limits are two standard deviations wide and are centered on the forecast values, <PRDN REV>. The second argument specifies the type of standard deviations to be calculated; valid types include ARFS for autoregressive models and RFSD for other regression models.

`CLIM` can also generate confidence limits for forecast errors, and cumulative confidence limits.

The optional-output command `$CLP` can then be specified to generate a confidence limit plot (upper and lower confidence limits versus time), or `$CDP` can be specified to generate an augmented confidence limit plot (upper and lower confidence limits and data versus time).

```
> $clp
TSPAK  / TSPLT

   ------------¦----------------------
  ¦                            o   o o¦
  ¦                              o    ¦
  ¦                                   ¦
  ¦    o     o   o o o     o o +   + +¦
  ¦        o           o         +    ¦
60-                                   -
  ¦      o   +   +   +   o + + X   X  ¦
  ¦o o +           +   +         X   X¦
  ¦        +   o                      ¦
  ¦          X   X   X     X          ¦
  ¦+   X +         X     +   X        ¦
  ¦  +     X   +       X              ¦
50-                                   -
  ¦X     X               X            ¦
  ¦  X         X                      ¦
   ------------¦----------------------
               1981

  1 + PRDNREV   2 X LCL REV   3 O UCL REV
  VS  TIME
  JUL 1980-DEC 1981
```

TIME SERIES DATA MANAGEMENT

The time series data management facility includes specific commands to retrieve time series from a file (**LOAD**); enter, update and modify time series data (**ENTER**, **REPLACE**); and write time series to a file (**FILE**).

The Active Series List

During an ISX session, the names of all time series retrieved via **LOAD**, entered or updated by **ENTER**, or updated or modified by **REPLACE** are maintained on a temporary list called the *active series list*. The **FILE** command writes all series currently on the active series list to a permanent file.

The active series list is particularly useful for periodically retrieving a related set of time series, updating each series, and then rewriting the set of updated series back to the original file.

The series most recently referenced by **ENTER** or **REPLACE** becomes the *current series.* The current series can be referred to in abbreviated form (*, rather than series name) in subsequent **ENTER** and **REPLACE** commands.

The **SERIES** command provides "housekeeping" operations associated with the time series data management facility, for example, listing, clearing, and resetting the active series list. The command:

```
> series
```

prints the active series list and the name of the current series. If the keyword T is included, the date range of each series is also printed:

```
> series t
```

The RENAME argument renames a series and adjusts the active series list accordingly. For example, to rename <DATA REV> to <DATA RVNU>:

```
> series rename data rev data rvnu
```

The CLEAR argument deletes **FILE**d and active series from core and resets the active series list:

```
> series clear
```

The RESET argument defines a new active series list. For example,

```
> series reset data rev data msg
```

resets the list to include only <DATA REV> and <DATA MSG>.

Reading Time Series Data into ISX

The **LOAD** command loads (reads) all time series saved in a file into core:

```
LOAD filename
```

(When specified in VM/CMS, only the filename is required. The default filetype and filemode are DATA and *, respectively.) The file should contain only time series and their time control blocks. `LOAD` can load files previously written by `FILE`. The list of all series loaded becomes the active series list. Because of the overhead involved in maintaining the active series list, `LOAD` should be used for data management purposes only. Use the more efficient `INPUT` command for routine data input.

Entering, Updating, and Modifying Time Series

The `ENTER` command, to enter and update time series, and the `REPLACE` command, to update or modify time series, have been discussed in the sections on creating and updating time series earlier in this chapter. We will simply note here that specifying a series identifier in either of these commands adds that series to the active series list and makes it the current series.

Writing Time Series to a File

The `FILE` command writes all series currently on the active series list to a file and resets the active series list:

```
FILE filename
```

If filename is not specified, it defaults to the file most recently `LOAD`ed or `FILE`d. (When specified on VM/CMS, only the filename is required. The default filetype and filemode are DATA and *, respectively.)

If the file does not exist, `FILE` creates it. If it does exist, `FILE` appends all series in the active series list to it. If the REPLACE argument is included:

```
FILE REPLACE filename
```

the current contents of the existing file are replaced by the series in the active series list.

As per the `OUTPUT` command, the FORMAT argument can be included to generate a listable file in place of the more efficient, default machine-readable format:

```
FILE FORMAT AUTO filename
```

INTERFACE BETWEEN TIME SERIES AND MATRIX PACKAGES

The `TSTOST` command can be used to create a two-way matrix from a set of time series. The time series become the variables in the matrix; the values over time become the observations.

The following example creates a matrix, MSER, with two variables, REVENUES and MESSAGES, containing the data from the time series <DATA REV> and <DATA MSG>:

```
> tstost time 1976 1 1980 12 mser $
> data rev vn revenues data msg vn messages
TSTOST: MATRIX MSER - REVENUES MESSAGES OBL
```

The matrix has 60 observations, reflecting the specified time interval January 1976 through December 1980. An additional variable, OBL, containing appropriate dates as observation labels, is automatically created in the new matrix.

The VN argument is used to supply *variable names* (REVENUES for <DATA REV>, MESSAGES for <DATA MSG>). If VN is not specified for a time series, the variable name is formed by concatenating its blocktype and blockname (e.g., DATAREV).

Once created, the new two-way matrix can be processed using the commands and routines designed for two-way matrix analysis that are discussed in the next chapter.

Similarly, STATLIB subroutines `TSTAB` and `TSTABK` can be used to generate tables from one or more series for use with TABLE, a STATLIB package of routines for managing and analyzing multiway tables.

CHAPTER 6

Two-Way Data Matrix Analysis

The ISX system provides many procedures for the management, display, and analysis of variables within two-way data matrices. This chapter reviews the data matrix format in ISX and describes the application of ISX commands frequently used in such analysis.

We begin with a typical ISX terminal session that performs a standard data analysis on some variables in a two-way data matrix.

A SAMPLE ISX SESSION

The following example demonstrates the sequence of interactive commands that might be used to begin an analysis of the relationship among residence extension development, median income, and household size for given central offices within a hypothetical telephone company. The variables EXTDEV (extension development), MEDINC (median income), and HHSIZE (household size) are in a matrix called DEMO, which is stored in a file called SAMPLE DATA A. This matrix and its variables will be referred to throughout this chapter.

The intent at this point is not to dwell on the details of each statement, or to develop the "best" statistical model, but instead to illustrate the interactive nature and sequential ordering of ISX processing. Each statement is entered and executed immediately. A command might be issued to create a new variable in core that can be used in later steps, or to use the results from one command to determine the next logical step in the analysis. Each ISX statement can be thought of as a building block in the analysis sequence.

```
> term
> input sample
INPUT FROM SAMPLE DATA * ...
  DYNINP: 28 *DAT DEMO
  DYNINP: 28 DATA DEMO
> prtst obs 1 5 demo
STPAK / STPRT

 MATRIX=TMP.    OBS    1-    5

OBL         EXTDEV   MEDINC    HHSIZE


MAIN          70.50 20394.00       3.10
CENTRAL       98.60 40391.00       4.20
HOLMDEL      131.90 92560.00       3.00
BRIGHTON     103.50 65829.00       2.50
EAST         179.60 85300.00       5.30
```

```
> plotst demo extdev : medinc hhsize
STPAK  / STPLT

         |------------|------------|------------|--------
        |                                          C     |
        |                                         C      |
        |                                  C             |
        |                                    C           |
     150-                                                -
        |                                                |
        |                                               C|
        |                                C               |
        |                                                |
        |                           C     C              |
     100-             C   C            C                 -
        |                                                |
        |               C                                |
        |          C                                     |
        |C                                               |
         |------------|------------|------------|--------
         20000        40000        60000        80000

        2  EXTDEV    VS  3 MEDINC

STPAK  / STPLT

         ----|------|------|------|------|------|----
        |                                           C|
        |                                          C |
        |                      C                     |
        |                                        C   |
     150-                                            -
        |                                            |
        |           C                                |
        |    C                                       |
        |                                            |
        |C    C                                      |
     100-    C                       C             C -
        |                                            |
        |                          C                 |
        |                                        C   |
        |            C                               |
         ----|------|------|------|------|------|----
            2.5     3     3.5     4     4.5     5

        2  EXTDEV    VS  4 HHSIZE
```

```
> corrst demo
REGPAK / REGCOR


  CORRELATION MATRIX:

                       1          2          3          4

  1 OBL          0.0

  2 EXTDEV       0.0        1.0000

  3 MEDINC       0.0        0.8283     1.0000

  4 HHSIZE       0.0        0.3665     0.1212     1.0000
```

```
> regst demo extdev medinc hhsize
REGST:

REGPAK / REGANL


  SAMPLE SIZE . . . . . . . .   15
  SUM OF WEIGHTS . . . . . . .  1.5000D+01
  ESTIMATED STD DEV . . . . .   1.9355D+01
  R SQUARED . . . . . . . . .   0.7579


  VARIABLE      COEFFICIENT     ESTD STD DEV        T

   0              6.8097D-01     2.1861D+01      0.0311
   3 MEDINC       1.4111D-03     2.5382D-04      5.5596
   4 HHSIZE       8.2343D+00     4.3627D+00      1.8874

   2 EXTDEV     DEPENDENT VARIABLE


  ANALYSIS OF VARIANCE

     SOURCE        DF        SS            MS             F

     REGRESSION     2     1.4074D+04    7.0370D+03     18.784
     ERROR         12     4.4954D+03    3.7462D+02

     TOTAL         14     1.8570D+04

REGST - RSID PLOT..$RP ..../IND1 IND2 ..
- RSID VS PRDN PLOT..$RVP, RSID STD NORMAL PLOT..$NRM
- CONF LIMIT PLOT..$CLP, RSID ACOR PLOT..$AP
- PRTL RSID PLOT..$PRP IND, PRTL ADJ RSID PLOT..$PAR IND
- REGR COEF CORR..$RCC, DURBIN-WATSON..$DW
```

```
> $rvp
STPAK  / STPLT

    --|--------|--------|--------|--------|--------|--
   |                                   C              |
   |                                                  |
 20-                                               C  -
   |C                                                C|
   |              C                           C       |
   |                 C           C                    |
  0-----------------------C----------------------------
   |                                                  |
   |                          C  C                    |
   |                  C                               |
-20-                 C                                -
   |                                             C    |
   |                                                  |
   |                                     C            |
    --|--------|--------|--------|--------|--------|--
      60       80      100      120      140      160

        7   RSID      VS   6 PRDN
        MATRIX=DEMO
```

Data Matrix Structure

A data matrix is a collection of two-dimensional data. The two dimensions are usually *variable* and *observation.* An observation can represent a person, time, location, experimental condition, or any other relevant dimension.

DYNCOR Blocks

Each data matrix consists of two DYNCOR blocks: a *data block* containing the actual data values and an associated *control block* containing the names of the variables in the matrix. Each block is identified by two words: blocktype and blockname. Both blocks associated with a matrix are assigned the same one- to four-character blockname. This blockname is also referred to as the *matrix name* in ISX. The blocktype of the data block is DATA, while the blocktype of the control block is *DAT—For example:

<DATA DEMO> data block of matrix DEMO
<*DAT DEMO> control block (list of variable names)

For compactness, most ISX statements (procedure commands and STATLIB subroutines) require the specification of only the matrix name.

Control Information

The control block contains the the names of the variables in the associated matrix; these can be up to eight alphanumeric characters in length.

Variable name OBL is reserved for a variable containing observation labels. Each observation label may contain up to eight characters. When STPAK prints variables within a matrix, it includes the OBL variable to identify each observation.

CREATING A DATA MATRIX IN ISX

A data matrix can be created in the ISX environment in two ways: it can be entered interactively within ISX or retrieved from an existing file. Once defined in ISX, a data matrix remains in core for the duration of the ISX session unless modified or deleted. Any newly created data matrix or variables can be referred to in subsequent commands during the ISX session.

Entering a Data Matrix Interactively within ISX

The commands `STDIM` and `STPUT` can be used to create interactively a data matrix within ISX. `STDIM` creates the control block, while `STPUT` creates the data block. Each variable's values are entered in the data block through individual `STPUT` commands.

The following example creates data matrix DEMO with four variables, OBL, EXTDEV, ADJINC, and HHSIZE, and with five observations.

```
> stdim demo 4 obl extdev medinc hhsize
STDIM:   DEMO, 4 VBLS:
         OBL EXTDEV MEDINC HHSIZE
> stput demo obl 1 5 main central holmdel $
>   brighton east
STPUT:   DEMO, VBL=OBL, OBS 1-5
> stput demo extdev 1 5 70.5 98.6 131.9 103.5 179.6
STPUT:   DEMO, VBL=EXTDEV, OBS 1-5
> stput demo medinc 1 5 2394 40391 92560 65829 85300
STPUT:   DEMO, VBL=MEDINC, OBS 1-5
> stput demo hhsize 1 5 5.1 4.2 3.0 2.5 5.3
STPUT:   DEMO, VBL=HHSIZE, OBS 1-5
```

In this example, `STDIM` sets up a matrix called DEMO with the four specified variables. Each `STPUT` command inputs data to DEMO for a given variable for observations 1 through 5. For variable EXTDEV the values for observations 1 through 5 are 70.5, 98.6, 131.9, 103.5, and 179.6.

Reading Data Matrices from a File

The `INPUT` command can be used to retrieve data matrices stored in a data file and input them into ISX. A data file that contains data matrices (data and control blocks) can be created in the EDIT environment or generated from within ISX. (Chapter 3, which is on Data in ISX, includes a description of formats used in data files.)

A data file may contain many data matrices. The `INPUT` command can be used to copy the entire contents of a file into ISX:

```
> input sample
INPUT FROM SAMPLE DATA * ...
  DYNINP: 28 *DAT DEMO
  DYNINP: 28 DATA DEMO
```

or to retrieve selected DYNCOR blocks:

```
> input sample data demo
INPUT FROM SAMPLE DATA * ...
  DYNINP: 28 *DAT DEMO
  DYNINP: 28 DATA DEMO
```

Verification of blocks retrieved may be suppressed through `INPUT`'s NOPRINT keyword:

```
> input noprint sample
```

If a file contains matrix data that is not in the form of DYNCOR blocks, the STDIM and STREAD commands can be used to enter the data into ISX. STDIM is used to create the control block, and STREAD reads the data from the file. The following example reads in data from a file to create a matrix, DEMO, with three variables, OBL, EXTDEV, and MEDINC, and five observations. The original data file appears as follows:

```
MAIN 70.5 2394
CENTRAL 98.6 40391
HOLMDEL 131.9 92560
BRIGHTON 103.5 65829
EAST 179.6 85300
```

The file may be read in using the commands:

```
> stdim demo 3 obl extdev medinc
STDIM:   DEMO, 3 VBLS:
         OBL EXTDEV MEDINC
> stread demo 10 (*) 5
STREAD: DEMO (3 BY 5), UNIT=10, 3 VBLS READ
```

Here "10" in the STREAD command refers to FORTRAN unit 10 that has been previously "connected," using an appropriate operating system command, to the data file. The "*" refers to reading the data in free format, wherein data values are separated by one or more blanks; otherwise a FORTRAN format can be used to read the data. The "5" refers to the five observations to be read from the file.

SAVING A DATA MATRIX

A data matrix that is created or modified during an ISX session will not be available once the ISX session is terminated. The new data matrix must be saved in a file if it is to be used in later sessions. The OUTPUT command can be used to record a copy of a data matrix in order to access it in subsequent ISX sessions.

The following example shows how to save a copy of a data matrix in ISX for future use:

```
> output newsamp data demo
OUTPUT (APPEND) TO NEWSAMP DATA ...
  DYNOUT: *DAT DEMO 28
  DYNOUT: DATA DEMO 28
```

In this example, the DYNCOR data matrix block, <DATA DEMO>, and its associated control block, <*DAT DEMO>, are output to a file called NEWSAMP (with, on VM/CMS, a default filetype of DATA and filemode of A).

Note that the control block is not specified in the `OUTPUT` command. It is saved automatically with the data block. If the file NEWSAMP DATA A did not exist, it would be created. The data are appended to the file, if it exists, unless the keyword REPLACE is specified before the filename, in which case any data existing in the file are replaced.

By default, the data are written to the file in machine readable format. This default format is recommended for compactness, efficiency, and accuracy, but the file cannot be listed or edited at the operating system level. To create a file that can be listed and edited, the FORMAT argument should be included in the `OUTPUT` command.

The keyword FORMAT with the option AUTO automatically determines a reasonable format for the data—For example:

```
> output format auto newsamp data demo
```

Or a user-specified FORTRAN format can be given. For example:

```
> output format r(6f9.1) newsamp data demo
```

generates a file containing a DYNCOR data block with six data values per line in F9.1 format.

`OUTPUT`'s FORMAT AUTO option cannot be used for data matrices containing OBL variables. If the FORMAT keyword is included, care must be exercised to format the non-numeric OBL variable properly. See the *Procedure Manual* for a complete description of `OUTPUT`'s capabilities.

DELETING DATA MATRICES FROM THE ISX ENVIRONMENT

Each data matrix occupies space in the dynamic core region of ISX. Data matrices that are no longer needed can be deleted, clearing space for other uses. The DELETE ST command can be used to delete all or specified matrices from core. For example:

```
> delete st
```

deletes all matrices (both data and control blocks) from core; the command

```
> delete st demo
```

deletes only the matrix DEMO.

LISTING DATA MATRIX AND VARIABLE NAMES

The NAMEST command can be used to list the names of all data matrices that exist within the current ISX session. For example:

```
> namest
 DEMO SAMP
```

lists the names of the two matrices currently in core, DEMO and SAMP.

If the NAMEST command is followed by matrix names, the variable names and number of observations for each matrix are listed. For example:

```
> namest demo samp
DEMO - 5 OBNS, 4 VBLS:
  OBL       EXTDEV    MEDINC    HHSIZE
SAMP - 5 OBNS, 2 VBLS:
  INWARD    OUTWARD
```

CHANGING DATA VALUES IN A DATA MATRIX

Data values for existing observations in a matrix can be changed and new ones can be added using the STPUT command.

In the following example, STPUT changes the third through the fourth observations of variable EXTDEV in matrix DEMO to 361 and 365, respectively:

```
> stput demo extdev 3 4 361 365
STPUT:   DEMO, VBL=EXTDEV, OBS 3-4
```

Any modifications made to a data matrix exist only in the copy of the matrix within the ISX workspace. The modified matrix must be saved in a permanent file if it is to be accessed in a later ISX session. The matrix can be saved using the `OUTPUT` command.

PRINTING A DATA MATRIX

The `PRTST` command can be used to print some or all of the variables within a data matrix. For example:

```
> prtst demo
STPAK / STPRT

 MATRIX=DEMO    OBS    1-    5

OBL          EXTDEV   MEDINC    HHSIZE

MAIN          70.50 20394.00      3.10
CENTRAL       98.60 40391.00      4.20
HOLMDEL      131.90 92560.00      3.00
BRIGHTON     103.50 65829.00      2.50
EAST         179.60 85300.00      5.30
```

prints all variables for all observations in matrix DEMO;

```
> prtst demo extdev
STPAK / STPRT

 MATRIX=DEMO    OBS    1-    5

OBL          EXTDEV

MAIN          70.50
CENTRAL       98.60
HOLMDEL      131.90
BRIGHTON     103.50
EAST         179.60
```

prints all observations for variable EXTDEV. Note that if variable OBL exists in the matrix, `PRTST` automatically prints it.

A user-specified format for printing can be included in the PRTST command. For example:

```
> prtst format f7.0 demo extdev
STPAK / STPRT

 MATRIX=DEMO    OBS    1-    5

OBL         EXTDEV

MAIN           71.
CENTRAL        99.
HOLMDEL       132.
BRIGHTON      104.
EAST          180.
```

prints variable EXTDEV, using format F7.0, and variable OBL for all observations in matrix DEMO.

To specify only a subset of observations to be printed, the argument OBS followed by the beginning and ending observations should be included in the PRTST command. For example:

```
> prtst obs 1 3 demo extdev
```

prints observations 1 through 3 of variable EXTDEV using the default format (F9.2).

VARIABLE MANIPULATION

When analyzing data it is often necessary to create new variables in a data matrix. ISX provides capabilities for performing arithmetic transformations, recoding observations, calculating constant functions, and creating indicator variables.

This section describes the commands and STATLIB subroutines most often used to perform transformations and manipulate variables. When new variables are created, the data matrix's control block is adjusted appropriately to account for the additional variables. The matrix with new, added variables exists only within the current ISX environment and must be stored in a permanent file if access during a later session is desired.

Arithmetic Transformations

The SETST command performs multiple transformations with one or more variables (and/or constants) within the same matrix and stores the result as a new variable within the matrix. The new variable, which will replace an existing variable with the same name, is calculated as the result of an observation-by-observation evaluation of an algebraic expression.

The following example illustrates the use of SETST to create a new variable, INCPERHH, which is equal to the ratio of MEDINC (median income) to HHSIZE (household size) in matrix DEMO:

```
> setst demo incperhh = medinc / hhsize
STFN2:   DEMO, DTA.0=DIV(MEDINC,HHSIZE)

STENL:   DEMO, 1 VBLS:
         INCPERHH

STFN1:   DEMO, INCPERHH=EQL(DTA.0)
```

Printing these three variables from matrix DEMO gives the following result:

```
> prtst demo medinc hhsize incperhh
STPAK / STPRT

 MATRIX=DEMO    OBS    1-    5

OBL         MEDINC    HHSIZE    INCPERHH

MAIN        20394.00     3.10   6578.71
CENTRAL     40391.00     4.20   9616.90
HOLMDEL     92560.00     3.00  30853.33
BRIGHTON    65829.00     2.50  26331.60
EAST        85300.00     5.30  16094.34
```

The square root of a variable can be calculated:

```
> setst demo sqextdev = sqr ( extdev )
STFN1:   DEMO, DTA.1=EQL(EXTDEV)

STFN1:   DEMO, DTA.0=SQR(DTA.1)

STENL:   DEMO, 1 VBLS:
         SQEXTDEV

STFN1:   DEMO, SQEXTDEV=EQL(DTA.0)
```

SETST is a procedure command that invokes the STATLIB routines STFN0, STFN1, and STFN2. These routines can also be executed directly from ISX, providing access to an extensive list of additional functions.

Functions over Observations

STATLIB subroutine STFNOB can be used to calculate a statistic or evaluate a function over all observations of a variable in a data matrix. The result of this calculation is printed in a one-line message and stored temporarily in ISX variable $XXX. The available functions and statistics are given in the following table:

Function	Result
AVG	Average
EXT	Extract *C*th element
MAX	Maximum
MED	Median
MIN	Minimum
ORD	*C*th order statistic
SSQ	Sum of squares
STD	Standard deviation
SUM	Sum
VAR	Variance

Some examples follow.

```
> stfnob demo sum extdev
STFNOB: DEMO, SUM(EXTDEV)=584.1
```

calculates the sum of the values of variable EXTDEV in matrix DEMO. This total value is stored in ISX variable $XXX and can be printed and/or used in subsequent commands. For example:

```
> / $xxx
584.1
> setst demo ptextdev = ( extdev / $xxx ) * 100
STFN1:   DEMO, DTA.1=DIV(EXTDEV,584.1)

STFN1:   DEMO, DTA.0=MLT(DTA.1,100)

STENL:   DEMO, 1 VBLS:
         PTEXTDEV

STFN1:   DEMO, PTEXTDEV=EQL(DTA.0)
```

If the total value is to be used in later calculations in the current ISX session, it would be wise to save the temporary contents of $XXX in another ISX variable, since ISX might subsequently redefine $XXX. For example:

```
> /def $tot $xxx
```

stores the total value in ISX variable $TOT.

When using the EXT and ORD functions, the number of the observation to be extracted and the fractional order statistic (between 0 and 1) should be given, respectively, after the variable name in STFNOB. For example:

```
> stfnob demo ext medinc 4
STFNOB: DEMO, EXT(MEDINC)=65829, ARG=4
```

extracts the fourth observation of variable MEDINC in matrix DEMO. A number greater than the number of observations in the data matrix translates to the last observation.

```
> stfnob demo ord extdev .25
STFNOB: DEMO, ORD(EXTDEV)=89.23333, ARG=.25
```

calculates the twenty-fifth fractional order statistic for variable EXTDEV in matrix DEMO.

Additional Capabilities

Indicator variables, useful as "dummy" variables in regression models, can be created from a variable in a data matrix through the `INDICAT` command. `STRCOD` recodes observations, that is, maps ranges of values of a variable into new values. `STFNC` computes functions over clusters (groups) of contiguous observations. These commands are beyond the scope of this primer.

MANIPULATING DATA MATRICES

It is often necessary to manipulate a data matrix when analyzing its variables. ISX provides capabilities for sorting a matrix, renaming a variable within a matrix, copying a matrix to a new matrix, extracting variables, and linking matrices.

When manipulating a matrix in the ISX environment, the control block is adjusted appropriately. Remember, the manipulated matrix exists only within the current ISX environment and must be stored in a permanent file if access will be desired at a later time.

Sorting a Data Matrix

The `STSORT` command sorts matrix observations so that specified variables are arranged in either ascending or descending order. For example:

```
> stsort demo 1 extdev a
STSORT: DEMO, DIR=A, VBLS:
        EXTDEV
> prtst demo
STPAK / STPRT

 MATRIX=DEMO    OBS    1-    5

OBL          EXTDEV    MEDINC     HHSIZE

MAIN           70.50 20394.00       3.10
CENTRAL        98.60 40391.00       4.20
BRIGHTON      103.50 65829.00       2.50
HOLMDEL       131.90 92560.00       3.00
EAST          179.60 85300.00       5.30
```

sorts all observations in matrix DEMO so that variable EXTDEV is arranged in ascending order. To sort in descending order, "d" should be used in place of "a" as the last argument.

STSORT sorts the matrix based on the number of variables specified in the second argument (1 in the preceding example) and listed thereafter. The first variable listed determines the primary sort, and so on.

Renaming Variables within a Data Matrix

The STNAM command renames one or more variables within a data matrix. The command requires the number of variables to be renamed, followed by pairs of old and new variable names. For example:

```
> namest demo
DEMO - 5 OBNS, 4 VBLS:
  OBL        EXTDEV    MEDINC     HHSIZE
> stnam demo 2 extdev nextdev hhsize nhhsize
STNAM:  DEMO, 2 OLD/NEW VBLS:
        EXTDEV NEXTDEV HHSIZE NHHSIZE
> namest demo
DEMO - 5 OBNS, 4 VBLS:
  OBL        NEXTDEV   MEDINC     NHHSIZE
```

lists DEMO's variable names, renames two variables, EXTDEV to NEXTDEV and HHSIZE to NHHSIZE, and then again lists the variable names.

Copying a Data Matrix

It is often desirable to copy all or selected observations from one matrix into another matrix. The `STCPY` command can be used in the ISX environment to accomplish this task. This command copies to a new matrix those observations with a nonzero value of a "copy variable." For example:

```
> prtst demo extdev medinc hhsize ind
STPAK / STPRT

 MATRIX=DEMO    OBS    1-    5

OBL         EXTDEV   MEDINC    HHSIZE   IND


MAIN          70.50 20394.00      3.10      1.00
CENTRAL       98.60 40391.00      4.20      1.00
HOLMDEL      131.90 92560.00      3.00      0.0
BRIGHTON     103.50 65829.00      2.50      1.00
EAST         179.60 85300.00      5.30      1.00

> stcpy demo ind ndem
STCPY:  DEMO TO NDEM, 4 OBS, CPY VBL=IND
> prtst ndem extdev medinc hhsize ind
STPAK / STPRT

 MATRIX=NDEM    OBS    1-    4

OBL         EXTDEV   MEDINC    HHSIZE   IND


MAIN          70.50 20394.00      3.10      1.00
CENTRAL       98.60 40391.00      4.20      1.00
BRIGHTON     103.50 65829.00      2.50      1.00
EAST         179.60 85300.00      5.30      1.00
```

In this example, all observations in matrix DEMO for which IND is nonzero are copied to NDEM.

`STCPY` copies the entire matrix when "...." is specified for the copy variable. If the same name is used for both the original and copied matrix, the copied data matrix overwrites the existing matrix.

The data transformation commands `STFN1` and `STFN2` can be used to create copying indicator variables.

Extracting Variables from a Data Matrix

The `STEXT` command extracts selected variables from a data matrix and stores them in a new matrix. For example:

```
> namest demo
DEMO - 5 OBNS, 4 VBLS:
  OBL       EXTDEV    MEDINC    HHSIZE
> stext demo dema 2 extdev medinc
STEXT:  DEMO TO DEMA, 2 VBLS:
        EXTDEV MEDINC
> namest demo dema
DEMO - 5 OBNS, 4 VBLS:
  OBL       EXTDEV    MEDINC    HHSIZE
DEMA - 5 OBNS, 2 VBLS:
  EXTDEV    MEDINC
```

lists DEMO's variable names and extracts variables EXTDEV and MEDINC from DEMO, storing them in matrix DEMA. The variable names for both DEMO and DEMA are then listed.

The new data matrix overwrites the old matrix if both have the same name.

Linking Data Matrices

`STLINK` links, or joins, two data matrices that have equal numbers of observations. The first matrix is enlarged to include the variables in the second matrix. When variables with the same name exist in both matrices, the contents of the variable in the second matrix replace the contents of that variable in the first matrix.

The following example illustrates the use of `STLINK` and shows the results of the linking process.

```
> namest demo dem1
DEMO - 5 OBNS, 4 VBLS:
  OBL       EXTDEV   MEDINC    HHSIZE
DEM1 - 5 OBNS, 2 VBLS:
  EXTDEV   SIZE
> prtst demo
STPAK / STPRT

 MATRIX=DEMO   OBS   1-   5

OBL        EXTDEV   MEDINC   HHSIZE

MAIN         70.50 20394.00      3.10
CENTRAL      98.60 40391.00      4.20
HOLMDEL     131.90 92560.00      3.00
BRIGHTON    103.50 65829.00      2.50
EAST        179.60 85300.00      5.30

> prtst dem1
STPAK / STPRT

 MATRIX=DEM1   OBS   1-   5

 EXTDEV   SIZE

    40.50     3.00
    67.80     7.00
    47.90     6.00
    75.30     8.00
    98.30     4.00

> stlink demo dem1
STENL:  DEMO, 2 VBLS:
        EXTDEV SIZE

STLINK: DEMO AND DEM1
```

```
> prtst demo
STPAK / STPRT

 MATRIX=DEMO    OBS    1-    5

OBL         EXTDEV    MEDINC    HHSIZE    SIZE

MAIN          40.50 20394.00      3.10      3.00
CENTRAL       67.80 40391.00      4.20      7.00
HOLMDEL       47.90 92560.00      3.00      6.00
BRIGHTON      75.30 65829.00      2.50      8.00
EAST          98.30 85300.00      5.30      4.00
```

In this example, variables EXTDEV and SIZE from matrix DEM1 are linked to matrix DEMO. The values of EXTDEV in DEM1 have replaced those in DEMO.

A data matrix can also be enlarged by linking observations using the `DYNCON` command. This command is beyond the scope of this primer but can be found in the *STATLIB Manual.*

PLOTTING VARIABLES IN A DATA MATRIX

There are many capabilities within the ISX environment for graphically displaying variables. With the `PLOTST` command, users can generate *scatter plots* of one or more variables versus another variable or versus observation order. Also, *box plots* and *histograms* are available for univariate displays of variables. By default, plots are printed at the user's terminal. The GRPAK graphics package, described in Chapter 9, provides additional graphics capabilities. By resetting several ISX parameters, graphical output can be directed to various screen and pen plotting devices.

Scatter Plots

`PLOTST` generates separate scatter plots of one or more variables versus another. Many different plots can be specified with one call to `PLOTST`.

An example of a simple scatter plot of one vertical variable versus one horizontal variable follows:

```
> plotst demo extdev : medinc
STPAK  / STPLT

    |-----------|-----------|-----------|-------
   |                                      C     |
   |                                     C      |
   |                               C            |
   |                                 C          |
150-                                            -
   |                                            |
   |                                           C|
   |                             C              |
   |                                            |
   |                        C     C             |
100-            C  C           C                -
   |                                            |
   |             C                              |
   |         C                                  |
   |C                                           |
    |-----------|-----------|-----------|-------
    20000       40000       60000       80000

   2   EXTDEV    VS   3 MEDINC
```

In this example, variable EXTDEV is plotted on the vertical axis and MEDINC on the horizontal axis. Both variables are from matrix DEMO.

The next example shows how to plot more than one vertical variable versus one horizontal axis variable:

```
> plotst demo medinc & nhhsize : extdev
STPAK   / STPLT

      ----|--------|--------|--------|--------|--------|
     |                           +                      |
     |                                                 +|
80000-  X           X                        X        + -
     |                  +                    +  +       |
     |    X       X ++         +                        |
60000-               +                          X       -
     |                                                  |
     |X          +               X                      |
40000-    +      X+  X  X      X                        -
     |  +            X                                  |
     |                                                  |
20000-+                                                 -
      ----|--------|--------|--------|--------|--------|
          80       100      120      140      160

     3 + MEDINC     5 X NHHSIZE     VS   2 EXTDEV
```

Here variables MEDINC (+) and NHHSIZE (X) are plotted on the vertical axis against EXTDEV, the horizontal axis variable. All variables are from matrix DEMO.

Variables can also be plotted against time (observation order) by specifying "...." as the horizontal axis variable. For example:

```
> plotst demo extdev : medinc ....
```

generates two separate plots: EXTDEV versus MEDINC and EXTDEV versus observation order.

More than one set of vertical variables can be specified before the colon, and more than one horizontal variable can be specified after the colon in the `PLOTST` command. Of course, this results in the generation of more plots. The number of plots generated is the number of sets of vertical variables times the number of horizontal variables specified in one `PLOTST` call. An exception is a plot of a variable against itself, which is always suppressed. The following example illustrates the specification of four plots in one command.

```
> plotst demo extdev & medinc medinc : .... hhsize
```

The four plots generated are:

1. EXTDEV and MEDINC versus observation order
2. EXTDEV and MEDINC versus HHSIZE
3. MEDINC versus observation order
4. MEDINC versus HHSIZE

If an OBL variable exists in the data matrix, its first character will be used as the plotting character when `PLOTST` plots a single variable on the vertical axis.

Box Plots

The `BOXPLT` command generates *box plots* for specified variables in a data matrix. This plot graphically displays the minimum, maximum, median, and the first and third quartiles of a variable. An example follows:

```
> boxplt demo extdev nhhsize
STPAK  / STPLT

    ------------!-----------------------!------------
   !            :                                    !
   !            :                                    !
   !            :                                    !
   !            :                       :            !
150-            -                       -            -
   !                                                 !
   !                                                 !
   !                                                 !
   !                                    X            !
   !            X                                    !
100-            -                                    -
   !            :                                    !
   !            :                                    !
   !            :                       -            !
   !            :                       :            !
    ------------!-----------------------!------------
                1                       2

      3 X MEDIAN     4 - Q25        2 - Q75
      5 : MINIMUM    1 : MAXIMUM    VS  6 VAR
```

The plots above the x-axis labels "1" and "2" refer to variables EXTDEV and NHHSIZE, respectively.

The optional keyword STATS, specified before the matrix name in the `BOXPLT` command, generates a listing of each variable's minimum, maximum, median, and first and third quartile values.

When GRPAK plotting (Chapter 9) is enabled, `BOXPLT` generates high-quality box plots on a graphical output device.

Histograms

The `HIST` command generates a separate *histogram* for each selected variable in a data matrix. If no variables are specified, a separate histogram is generated for each variable in the matrix. In the following example a histogram is generated for variable HHSIZE in matrix DEMO:

```
> hist demo hhsize
STPAK  / STPLT    HISTOGRAM FOR HHSIZE

  ----¦------¦------¦------¦------¦------¦----
 ¦                                          + ¦
 ¦                                          + ¦
 ¦                                          + ¦
4-    +                                     + -
 ¦    +                                     + ¦
 ¦    +                                     + ¦
 ¦    +                                     + ¦
 ¦    +                                     + ¦
2-    +                      +              + -
 ¦    +                      +              + ¦
 ¦ +  +     +  +        +    +              + ¦
 ¦ +  +     +  +        +    +              + ¦
 ¦ +  +     +  +        +    +              + ¦
0- +  +     +  +        +    +              + -
  ----¦------¦------¦------¦------¦------¦----
      2.5    3      3.5    4      4.5    5

 2 + TOTAL     VS   1 HHSIZE
```

The histogram plotted has a default of fifteen intervals. The number of intervals can be specified by using the optional keyword INTRVLS, followed by the desired number, before the data matrix name—For example:

```
> hist intrvls 7 demo hhsize
STPAK  / STPLT    HISTOGRAM FOR HHSIZE

  ----|------|------|------|------|------|----
 |   +                                    +   |
 |   +                                    +   |
 |   +                                    +   |
4-   +                                    +   -
 |   +                                    +   |
 |   +                                    +   |
 |   +                                    +   |
 |   +                                    +   |
2-   +                        +           +   -
 |   +                        +           +   |
 |   +     +     +      +     +           +   |
 |   +     +     +      +     +           +   |
 |   +     +     +      +     +           +   |
0-   +     +     +      +     +           +   -
  ----|------|------|------|------|------|----
      2.5    3      3.5    4      4.5    5

 2 + TOTAL      VS   1 HHSIZE
```

When GRPAK plotting (Chapter 9) is enabled, `HIST` produces high-quality histograms on a graphical output device.

DISPLAYING DESCRIPTIVE STATISTICS

`DSTATST` prints the mean, median, standard deviation, and other descriptive statistics for specified variables in a data matrix. If no variables are specified, the statistics are calculated for all variables in the matrix. An example follows:

```
> dstatst demo extdev medinc
STPAK / STPRT

 MATRIX=TMP.    OBS    1-    2

OBL          MEAN        MEDIAN      TRIMMEAN MIDMEAN   HUBER

EXTDEV         118.89    105.00    117.95    114.83    118.89
MEDINC       61830.93 66289.00 62654.62 63332.11 61830.93

OBL          MINIMUM   Q25         Q75         MAXIMUM   VARIANCE

EXTDEV          70.50     96.68    153.90    179.60   1326.39
MEDINC       20394.00 43380.00 75077.17 92560.00*********

OBL          STD.DEV   MAD         UMAD

EXTDEV          36.42     24.50     36.32
MEDINC       20531.86 16921.00 25086.73

 15 OBSERVATIONS IN DEMO
```

In the example, descriptive statistics are calculated for both EXTDEV and MEDINC in matrix DEMO. The asterisks shown for the variance of MEDINC indicate that the value was too large to print with the default format of F9.2. A format for printing the descriptive statistics can be requested, as shown below:

```
> dstatst format f9.0 demo extdev medinc
```

CORRELATION COEFFICIENTS

CORRST prints the *correlation matrix* for specified variables in a data matrix. For example, the following calculates and prints the correlation matrix for all variables in matrix DEMO:

```
> corrst demo
REGPAK / REGCOR


   CORRELATION MATRIX:

                      1         2         3         4

   1 OBL          0.0

   2 EXTDEV       0.0       1.0000

   3 MEDINC       0.0       0.8283    1.0000

   4 HHSIZE       0.0       0.3665    0.1212    1.0000
```

In the following example, however, the correlation matrix is calculated and printed for only two variables, EXTDEV and MEDINC:

```
> corrst demo extdev medinc
REGPAK / REGCOR


   CORRELATION MATRIX:

                      1         2

   1 EXTDEV       1.0000

   2 MEDINC       0.8283    1.0000
```

If the data matrix contains a variable WGTS, the values in this variable are used to compute weighted correlation coefficients.

If the keyword DSP is included in the CORRST command, a symbolic *triangular display* is generated, in addition to the correlation matrix—For example:

```
> corrst dsp demo
REGPAK / REGCOR


  CORRELATION MATRIX:

                     1          2          3          4

  1 OBL          0.0

  2 EXTDEV       0.0        1.0000

  3 MEDINC       0.0        0.8283     1.0000

  4 HHSIZE       0.0        0.3665     0.1212     1.0000

TRIPAK / TRIDIS

   1 OBL
   2 EXTDEV      X
   3 MEDINC      X X
   4 HHSIZE      + + X

               1 2 3 4

  SYMBOLS: -(-.1,-.4) =(-.4,-.7) #(-.7,-1.) <(<-1.)
           +(+.1,+.4) *(+.4,+.7) X(+.7,+1.) >(>+1.)
```

The symbols in the triangular display correspond to the ranges of the correlations listed below the display. To obtain only the symbolic display, include the keyword NOPRINT.

REGRESSION ANALYSIS

The `REGST` command provides regression capabilities for two-way matrix data, including ordinary least squares, weighted least squares, and robust regression analysis.

An example of an *ordinary least squares* regression follows:

```
> regst demo extdev medinc hhsize
REGPAK / REGANL


  SAMPLE SIZE . . . . . . . . .     15
  SUM OF WEIGHTS . . . . . . . .   1.5000D+01
  ESTIMATED STD DEV  . . . . .    1.9355D+01
  R SQUARED  . . . . . . . . .    0.7579


  VARIABLE      COEFFICIENT      ESTD STD DEV        T

   0             6.8097D-01        2.1861D+01       0.0311
   3 MEDINC      1.4111D-03        2.5382D-04       5.5596
   4 HHSIZE      8.2343D+00        4.3627D+00       1.8874

   2 EXTDEV     DEPENDENT VARIABLE


  ANALYSIS OF VARIANCE

     SOURCE        DF        SS              MS              F

     REGRESSION     2     1.4074D+04     7.0370D+03      18.784
     ERROR         12     4.4954D+03     3.7462D+02

     TOTAL         14     1.8570D+04

REGST - RSID PLOT..$RP ..../IND1 IND2 ..
- RSID VS PRDN PLOT..$RVP, RSID STD NORMAL PLOT..$NRM
- CONF LIMIT PLOT..$CLP, RSID ACOR PLOT..$AP
- PRTL RSID PLOT..$PRP IND, PRTL ADJ RSID PLOT..$PAR IND
- REGR COEF CORR..$RCC, DURBIN-WATSON..$DW
```

This command calculates a regression model with dependent variable EXTDEV and independent variables MEDINC and HHSIZE, all from matrix DEMO. The regression output produced by `REGST` includes sample size, sum of weights used, estimated error standard deviation, R-squared, estimates of the regression coefficients along with their t-statistics, and an analysis of variance (ANOVA) table.

The optional-output list in the preceding example indicates various results and plots that can be generated after calling `REGST`. Optional output is requested by specifying the appropriate command. To generate a plot of residuals versus predicted values, specify `$RVP`:

```
> $rvp
STPAK  / STPLT

    --|--------|--------|--------|--------|--------|--
   |                                   C              |
   |                                                  |
 20-                                               C  -
   |C                                                C|
   |              C                           C       |
   |                 C           C                    |
  0-----------------------C----------------------------
   |                                                  |
   |                          C  C                    |
   |                  C                               |
-20-                 C                                -
   |                                             C    |
   |                                                  |
   |                                     C            |
    --|--------|--------|--------|--------|--------|--
      60       80      100      120      140      160

        7   RSID      VS  6 PRDN
        MATRIX=DEMO
```

To generate a standard normal plot of the regression residuals, specify **$NRM**:

```
> $nrm
STPAK  / STPLT    NORMAL PLOT OF RESIDUALS

         -----------|------------|------------|-----------
        !                                                X!
        !                                                +!
        !                                         X       !
      20-                                         +       -
        !                                 +   +           !
        !                            +  +                 !
        !                        + + X                    !
       0-                      + X                        -
        !                    X X                          !
        !                 +  +                            !
        !           X   +                                 !
     -20-           +                                     -
        !       +                                         !
        !                                                 !
        !+                                                !
         -----------|------------|------------|-----------
                   -1            0            1

        1 + RESIDUAL  3 X LINE      VS  2 NORMAL
```

The following table lists each command and the resulting output:

COMMAND	OUTPUT
`$RP ind1 ind2..`	Residuals vs independent variable plot
`$RVP`	Residuals vs predictions plot
`$NRM`	Standard normal plot of residuals
`$CLP`	Confidence limit plot for predictions
`$AP`	Autocorrelation plot for residuals
`$PRP ind`	Partial residual plot
`$PAR ind`	Partial adjusted residual plot
`$RCC`	Regression coefficient correlation matrix
`$DW`	Durbin-Watson statistic

For `$RP`, "...." may be listed as an independent variable to obtain a plot of residuals versus observation order.

Defining ISX variables with the same names as these commands should be avoided.

`REGST` also creates two new variables, PRDN (predicted values) and RSID (residuals). Both of these variables are added to the original matrix used in the regression.

To perform a *weighted regression*, a variable with name WGTS must be present in the data matrix (the STPUT command can be used to create new variables in a matrix).

REGST provides three options for *robust regression* analysis: HUBER, BISQUARE and ROBUST, where ROBUST uses a combination of the Huber and bisquare weighting techniques. An example using the ROBUST option follows:

```
> regst robust demo extdev medinc hhsize
```

Here REGST calculates a robust regression using the combined Huber-bisquare weighting scheme for the model EXTDEV = f(MEDINC,HHSIZE). See *The Beginning Forecaster* for an application of robust regression.

When called with one of the robust options, REGST creates and leaves in the matrix a variable called *RWGTS. This variable contains the final robust weights used.

An additional keyword, CORR, instructs REGST to print the correlation matrix in the course of estimating the regression coefficients.

INTERFACE BETWEEN MATRIX AND TIME SERIES PACKAGES

The STTOTS command creates time series from variables in a two-way data matrix. Each variable becomes a time series and its observations become values over time.

The following example illustrates the use of STTOTS to create time series from variables EXTDEV and MEDINC in matrix DEMO:

```
> sttots 1981 1 4 demo extdev tsi data extd $
>   medinc tsi data mein
STTOTS: DEMO ->
DATA EXTD  , DATA MEIN
```

The time series will be quarterly (4) and will start in the first quarter of 1981 (1981 1). The TSI argument is specified after each variable name to specify a time series identifier (blocktype and blockname) for that variable (<DATA EXTD> for EXTDEV, <DATA MEIN> for MEDINC). If TSI is not specified, the blocktype and blockname of the created time series will be the first four and last four characters in the variable name, respectively. If the variable has a name with four or fewer characters, the time series blocktype will be DATA and the blockname will be the variable name.

Once created, the new time series are available for use with the time series analysis commands, which were discussed in Chapter 5.

CHAPTER 7

Programming in ISX

ISX is a true programming language that enables one to "package" a group of commands and execute them by issuing a single command. Users might wish to do this to eliminate repetitive typing, or to construct simple (or sophisticated) ISX commands of their own.

This chapter first discusses execution of commands from files. Then the creation and use of user-defined commands, or procedures, is considered.

EXECUTING COMMANDS FROM FILES

ISX normally reads command lines from the terminal. But, with the `/XEQFILE` command, you can tell ISX to read and execute commands stored in a disk file. For example, the following file, named MONTHLY (with a filetype of ISX on VM/CMS), might be created using the system editor to perform the same analysis on a set of time series after it is updated each month.

```
INPUT REVENUE
SET DATA REV = DATA BREV + DATA RREV
PRTTS TIME 1972 3 9999 1 DATA REV
PLOTTS DATA REV : .... ; DATA MSG ; DATA HOUS
```

Note that we show file contents in upper case, without prompts (as they would appear if the file were printed), to distinguish them from commands entered from the terminal.

To execute the file, enter the ISX command

```
> /xeqf monthly
```

Data are then read from the file REVENUE. The series <DATA REV> is constructed by adding <DATA BREV> and <DATA RREV>, printed from March 1972 through its end (assumed to be before January 9999), and finally plotted against time, against <DATA MSG>, and against <DATA HOUS>. Control is then returned to the terminal.

HELPFUL ISX COMMANDS

When you execute sets of commands from a file, you will no longer have a record on your terminal output of the commands being executed—all you will see is the resulting output. Several ISX commands help to make the output more readable and help to trace the progress of the "program."

Command Verification

The `/VER` command requests that ISX print each command before executing it. To do this every time a command file is executed, simply insert the line

```
/VER
```

at the top of the file, and the line

```
/VER *
```

at the bottom (to reset verification to off). If you do not place `/VER` statements in the file, but wish to turn verification on while the file runs (perhaps to locate a problem), issue the commands

```
> /ver
> /xeqf monthly
> /ver *
```

from the terminal.

Printing Messages

STATLIB output is, to a large degree, self-documenting. Each of the commands in the MONTHLY file just given prints at least a short message indicating what it has done. `INPUT`, for example, prints the name of the file it is reading and the name of each block loaded. You may add your own messages, using the / command. In the following modified version of the MONTHLY file, the NOPRINT keyword suppresses `INPUT`'s messages. The file's second line supplies a simple one-line message confirming that the data are loaded.

```
INPUT NOPRINT REVENUE
/ REVENUE DATA LOADED.
SET DATA REV = DATA BREV + DATA RREV
PRTTS TIME 1972 3 9999 1 DATA REV
PLOTTS DATA REV : .... ; DATA MSG ; DATA HOUS
```

Comments

Comment lines are useful in documenting programs; they provide at least a few clues about the program that may be helpful to the programmer at some future time, as well as to other readers. In ISX, the `/COMMENT` command serves this purpose. The statement is essentially ignored by ISX—it is not printed when executed. `/COMMENT` can be abbreviated `/C`, as shown in this version of MONTHLY:

```
/C  MONTHLY REVENUE ANALYSIS .. JLB .. 5/7/82
/C  SHOULD BE RUN ON 3RD WORKING DAY OF EACH MONTH
INPUT NOPRINT REVENUE
/ REVENUE DATA LOADED.
SET DATA REV = DATA BREV + DATA RREV
PRTTS TIME 1972 3 9999 1 DATA REV
PLOTTS DATA REV : .... ; DATA MSG ; DATA HOUS
```

USER-DEFINED PROCEDURES

Command files, such as in the preceding example, are useful when you wish to always execute exactly the same commands. The ISX procedure capability allows you to create commands that, like other ISX commands, may have arguments passed to them. Several ISX commands are available to control the sequence of execution of such procedures.

Defining Procedures

In its simplest form, a procedure (or *proc*) contains a set of statements to be executed "as is." Procs are created by the /PROC statement:

```
> /proc revanl
>   prtts time 1972 3 9999 1 data rev
>   plotts data rev : .... ; data msg ; data hous
> /procend
```

The procedure name (here REVANL) may contain up to seven characters. Statements within the proc are indented two spaces to improve readability. The procedure definition is terminated with the /PROCEND command. REVANL is now a command that will print and plot <DATA REV>.

Procedure REVANL was entered into ISX from the terminal, and will be lost on termination of the ISX session. This may be satisfactory for short procedures that have limited use, but in most cases you will prefer to store procedure definitions in files and execute the file as needed. So, for example, REVANL might be placed, along with other procedure definitions and commands, in a file named MYPROCS. Executing the file (/xeqf myprocs) would, in effect, load REVANL into core so it could be used as a command.

Since files containing procedure definitions are typically executed immediately upon entry into ISX, some systems (such as VM/CMS) allow filenames to be passed as arguments to ISX:

```
. isx myprocs ourprocs initcmds
```

The files are then executed in reverse order.

Procedure Arguments

Procedure REVANL suffers from lack of generality: It always analyzes the same time series. One way to generalize it is to replace DATA REV by $SER and then define $SER as the desired series name before calling REVANL. ISX provides an easier way: *Symbolic names* can be included in the `/PROC` statement; occurrences of the names inside the proc are replaced by the corresponding arguments passed to the proc at execution—For example:

```
> /proc revanl type name
>   prtts time 1972 3 9999 1 type name
>   plotts type name : .... ; data msg ; data hous
> /procend
```

Here TYPE and NAME are symbolic names referring to the first and second arguments passed to REVANL. With this definition, the commands

```
> revanl data rev
> revanl data brev
> revanl data rrev
```

perform the analysis for <DATA REV>, then for <DATA BREV>, and finally for <DATA RREV>.

Another way to refer to arguments within a procedure is by *argument number*: $1 is the first argument, $2 the second, and so on. Thus, an equivalent version of REVANL is

```
> /proc revanl
>   prtts time 1972 3 9999 1 $1 $2
>   plotts $1 $2 : .... ; data msg ; data hous
> /procend
```

Reference by symbolic names is more convenient (and readable) in many situations (such as REVANL), and reference by argument number is useful in other cases.

ISX also provides a way to refer collectively to an argument and all succeeding arguments. Special variables of the form $*n*.. are used for this purpose, *n* being a variable number. If "$1 $2" is replaced by "$1.." in REVANL's `PRTTS` statement, as follows:

```
> prtts time 1972 3 9999 1 $1..
```

then all procedure arguments will be passed on to `PRTTS`, so that any number of time series can be printed.

Redefining Arguments

As discussed earlier, arguments are defined when the procedure is called. They can also be defined (or redefined) within the procedure via the `/ARGS` command:

```
> /args data prdn rsid
```

In this example, argument 1 ($1) is defined as DATA, argument 2 ($2) as PRDN, and argument 3 ($3) as RSID. All other arguments become undefined.

```
> /args $1 $2 msg
```

leaves arguments 1 and 2 unchanged and defines argument 3 as MSG, *if* arguments 1 and 2 are currently defined. If only one were defined, it would become argument 1 and MSG argument 2; if neither were defined, MSG would become argument 1.

Skipping Statements

Within a procedure, statements are normally executed in the order they are encountered. With the `/SKIP` command you can change the order of execution. The simplest case is shown in the following example:

```
> /proc revanl type name
>   prtts time 1972 3 9999 1 type name
>   /skip 1 u
>     plotts type name : .... ; data msg ; data hous
> /procend
```

Here we instruct ISX to skip one statement, PLOTTS; the argument "U" means unconditionally. A more interesting (and useful) example would skip the PLOTTS statement if the series has a type other than DATA:

```
> /proc revanl type name
>   prtts time 1972 3 9999 1 type name
>   /skip 1 ne type data
>     plotts type name : .... ; data msg ; data hous
> /procend
```

When the second argument to /SKIP is EQ or NE, the skip occurs if the following two arguments are equal or not equal, respectively. If one of the two is missing, they are not equal; if both are missing, they are equal. Thus,

```
> /skip -2 ne $1
```

skips to the second preceding line if argument 1 is not null. An optional fifth argument to /SKIP tells ISX how many characters to compare; normally all characters must match.

Numerical comparisons may be made using codes LT, LE, GT and GE. For example,

```
> /skip 2 gt $3 1.4
```

skips two lines if $3 is greater than 1.4.

Looping

The ability to redefine arguments and alter the flow of control are the basic tools needed to program in ISX. The following example generalizes REVANL to a procedure that accepts as arguments an arbitrary number of time series blocktypes and blocknames, and processes each in turn:

```
> /proc revanl type name
>   prtts time 1972 3 9999 1 type name
>   plotts type name : .... ; data msg ; data hous
>   /args $3..
>   /skip -3 ne name
> /procend
```

The procedure loops, printing and plotting each series. After the last series is plotted, `/SKIP` fails because the second procedure argument, NAME, is null.

Skipping more than a few lines can be tedious and dangerous: tedious with respect to counting lines and dangerous with respect to errors in counts and failure to adjust counts when lines are added or removed at some later time. This problem is overcome by skipping to *labeled statements*. A label may be inserted at the beginning of any ISX statement within a procedure. Labels consist of one to three characters enclosed in angle brackets.

```
> /proc revanl type name
>   <top> prtts time 1972 3 9999 1 type name
>   plotts type name : .... ; data msg ; data hous
>   /args $3..
>   /skip <top> ne name
> /procend
```

Argument Shifting

In the preceding example, the statement

```
> /args $3..
```

is easier to write, and more certain to include all arguments, than a statement such as

```
> /args $3 $4 $5 $6 $7 $8 $9 $10 $11 $12
```

It is also more efficient computationally. An even more efficient way, equivalent to the former, is

```
> /argls 3
```

which performs a "left shift" on the arguments so that argument 3 becomes argument 1, argument 4 becomes argument 2, and so on.

Additional arguments may be included, so that

```
> /argls $1 $2 4
```

leaves the first two arguments untouched and shifts the remaining arguments by one.

Local Variables

The ISX variables we have encountered thus far have been *global variables*, that is, they can be set or used anywhere, inside or outside of procedures. For example, in the following version of REVANL, variable $TIM is used inside the procedure to hold some years and months temporarily, but it will still have those values after the procedure terminates.

```
> /proc revanl type name
>    /def $tim 1972 3 9999 1
>    /skip 2 ne $1 time
>      /def $tim $2 $3 $4 $5
>      /argls 6
>    <top> prtts time $tim type name
>    plotts type name : .... ; data msg ; data hous
>    /args $3..
>    /skip <top> ne name
> /procend
```

REVANL now has an optional TIME keyword. If it is present as the first argument, arguments 2 through 5 become the time period. Otherwise, "1972 3 9999 1" is used.

But the global variable $TIM might conflict with usage elsewhere of a variable with the same name. Further, if it is not deleted when the procedure terminates, it remains as clutter in the ISX workspace.

This problem is avoided by replacing $TIM with a *local variable*, say $.TI. Local variables have a period (.) as their first character and are known only within the current invocation of the procedure. The procedure may even call itself with no conflict. When the procedure completes, all local variables are deleted.

Local variables should be used whenever possible. If you need to use a global variable to return an answer, consider using one of the standard "temporary" variables ($XXX, $ERR, $TYP, $NAM, and $EOF), as appropriate.

Finishing Touches

As mentioned earlier in this chapter, comment lines are important in documenting software. When creating a file containing one or more procedures, it is good practice to place comments describing each procedure immediately preceding the /PROC statement. Additional comments can be placed inside the procedure if needed. Placement outside the procedure is preferred, however, since those comments are processed (ignored) when the file is executed, while comments inside a procedure remain there and are processed each time the procedure is executed (they are, however, reduced to minimal size).

The "final" version of REVANL, then, might be stored in a file with other procedures and appear as follows:

```
/C   MONTHLY REVENUE ANALYSIS .. JLB 5/7/82
/C   SHOULD BE RUN ON 3RD WORKING DAY OF EACH MONTH
/C
/C   USAGE:  REVANL [TIME N1Y N1M N2Y N2M] TYPE NAME ..
/C
/PROC REVANL TYPE NAME
  /DEF $.TI 1972 3 9999 1
  /SKIP 2 NE $1 TIME
    /DEF $.TI $2 $3 $4 $5
    /ARGLS 6
  / MONTHLY ANALYSIS FOR $1..
  / TIME PERIOD = $.TI
  <TOP> PRTTS TIME $.TI TYPE NAME
  PLOTTS TYPE NAME : .... ; DATA MSG ; DATA HOUS
  /ARGS $3..
  /SKIP <TOP> NE NAME
/PROCEND
```

Files of frequently used procedures may be stored and loaded more efficiently if they are processed by PROCDMP. See the *Procedure Manual* for details.

CHAPTER 8

Utility Routines

The STATLIB/ISX system provides substantial flexibility in controlling your terminal session, manipulating data, and performing statistical analysis. This chapter provides further information on STATLIB utility functions, including setting of print and execution options, directing output to files or printers, manipulating DYNCOR blocks, and executing operating system commands from within ISX.

SESSION CONTROL

ISX session control commands perform such housekeeping functions as setting output page width and depth, suppressing output or redirecting it to a file or line printer, changing plotting characters, and specifying plotting scales.

Specifying Page Size

While most STATLIB routines produce printed output that does not exceed 80 columns in width, some (such as the plotting and general-purpose printing routines) generate 132-column output. Since most terminals are limited to 80 columns, **TERM** (Chapter 2) sets the maximum width to 80.

If you have a wider terminal, say 120 columns, call **TERM** as follows:

```
> term * pagsiz 120
```

Here the asterisk means "for all routines." You can be more selective:

```
> term *prt pagsiz 120 *plt pagsiz 132
```

In this case the page size is set to 120 for all STATLIB routines whose names end in "PRT" (i.e., printing routines), and to 132 for plotting routines.

In addition to telling STATLIB what size lines to print, you may also need to tell the operating system. Most computer operating systems will fold output lines after a certain point, that is, print the remainder on the next line. To prevent this, tell the system your terminal is at least as wide as it really is.

Page depth (number of lines per page) is normally modified only for printer-plots. This is accomplished by specifying the number of lines desired *within* each plot, such as:

```
> term lines 25
```

In this example, the maximum size of plotted output is reduced to 25 lines from the default of 45. Note that only two arguments are specified: "lines" followed by the value; output from all plotting routines is affected.

Plotting Parameters

STATLIB plotted output (from commands such as PLOTTS and PLOTST) is generated by subroutine PPLOT, which prints printer-plots on the terminal or printer and/or calls GRPAK (Chapter 9) to create high-quality, high-resolution output on a graphical device. To modify the way plots are drawn, call PPLOT's parameter-setting command, PPLOTP. For example,

```
> pplotp char 3 123
```

changes the plotting characters for the first three variables on succeeding plots from the defaults "+," "X," and "O" to "1," "2," and "3." Scaling, which normally is adjusted so that all the data to be plotted just fits, may be controlled. For example,

```
> pplotp vmin 12.5
```

sets the vertical minimum for the next plot to 12.5. To draw a horizontal axis on the next plot at 1.12, the following command is issued:

```
> pplotp haxis 1.12 -
```

Here the minus sign is the plotting character to be used.

Similar `PPLOTP` parameters define plot mode (down the page rather than across), print a title above the plot, draw bars between variables or to a specified point, set tick-mark spacing, turn on grid lines, etc. Most parameters apply only to the next plot drawn.

The LINES parameter of `TERM` is actually a `PPLOTP` parameter, but, for reasons to be discussed shortly, is normally specified through `TERM`. Other `TERM` parameters that affect plots are PAGSIZ (discussed previously) and DEVICE. The output device is, by default, assumed to be a line printer (PRTR), and `PPLOT` draws a "box" around each plot. `TERM` sets the device to terminal (TERM), so that the time-consuming box is suppressed. If you want the boxes drawn on the terminal, issue

```
> term * device prtr
```

Sending Output to a Printer or File

STATLIB output may be directed to a high-speed line printer or into a disk file through the `OUTPR` command. `OUTPR`'s calling sequence is similar to `TERM`'s; you may specify various options to apply to the printed output. For example,

```
> outpr * pagsiz 68 lines 28
```

specifies that output is to be routed to the printer with page size limited to 68 and plot lines to 28. To return to terminal output, enter

```
> term
```

During a session, selected output can be routed to the printer by alternating calls to OUTPR and TERM. Each remembers arguments passed to it in previous calls, so no arguments are required. To close the printer file and release it for printing, enter

```
> clspr
```

OUTPR and CLSPR also accept various system-dependent options; these must be preceded by a semicolon. Such options might include number of copies and mailing address. One such option is FILE, which indicates that the output should be stored in a disk file (for later editing, printing, etc.) rather than being routed to a printer:

```
> outpr ; file tuesday
```

On VM/CMS, this sends the output to the file TUESDAY LISTING.

The following example summarizes the use of TERM and OUTPR.

```
> term * pagsiz 75
> input ..
> reg ..
> $rp
> outpr ; copies 2
> $rp
> term
> fcst ..
> clim ..
> $clp
> outpr
> $clp
> term
..
> clspr
> /stop
```

After entering ISX, TERM sets up terminal parameters; subsequently, calls to OUTPR and TERM direct output to the printer and terminal, respectively. Finally the print output is closed.

Suppressing Printing

STATLIB gives you the ability to suppress selectively printed output from subroutines or sets of subroutines. To stop printed output from all plotting routines, type

```
> option *plt print off
```

To suppress all TSPAK output, use

```
> option ts* print off
```

The asterisk, meaning "all," may be used either at the beginning or end of the first argument or, to suppress all printing, it may be used in the following manner:

```
> option * print off
```

To remove an entry in the OPTION table, for example, to turn TSPAK printing back on, replace off with an asterisk,

```
> option ts* print *
```

The OPTION table holds a limited number of entries, so unneeded ones should be removed.

`OPTION` is used by `TERM` and `OUTPR` to set the PAGSIZ and DEVICE options. The same option codes are recognized by all three, but `TERM` and `OUTPR` are normally used to set options that apply only to terminal or printer output. You can, for example, suppress printer output (but not terminal output) from STPAK routines:

```
> outpr st* print off
```

REGRESSION PARAMETER SETTING AND RESULT RETRIEVAL

STATLIB regression parameters are set through `REGPAR` (see *STATLIB Manual*, Chapter 10). The parameters most likely to be changed are those dealing with the presence or absence of a constant term and the collinearity threshold.

By default, regression models include a constant term. To suppress the constant in future regressions, enter:

```
> regpar con off
```

With the constant removed, the regression line is forced through the origin, and certain statistics, such as *F* and *R*-squared, have different interpretations than they do when the constant is present.

To turn the constant back on for future regression models, use:

```
> regpar con on
```

`REGGET` (see Chapter 10, *STATLIB Manual*) retrieves regression coefficients and other results, and stores them (up to 50) in an internal register called the *X* array. The `/DEFX` command retrieves information from the *X* array and places it in an ISX variable. The following commands regress the time series <DATA REV> on <DATA MSG> and <DATA HOUS>, and retrieve the two coefficients as $COE and the estimated standard deviation as $S2:

```
> reg rev ; msg ; hous
> regget b 1 2
> /defx $coe 2 r
> regget s2 1 1
> /defx $s2 1 r
> / depv = rev - msg & hous coeffs = $coe - s2 = $s2
```

In this example, `REGGET` first stored in the *X* array words 1 through 2 of *B* (the coefficients). Then `/DEFX` defined $COE as the first two real words of the *X* array. Similarly words "1 1," meaning 1 through 1, of *S2* (a scalar, it has only one word) were retrieved and defined as $S2. Finally, the summary line was printed.

Other information available through `REGGET` includes variable averages and standard deviations, correlations, sums of squares, number of variables, and sum of weights.

WORKSPACE (DYNCOR) CONTROL

Chapter 3 described STATLIB's dynamic core storage system, DYNCOR, which manages blocks of data and control information. DYNCOR usually operates quietly, with little or no user intervention, moving blocks when necessary to accommodate new ones. Sometimes, however, the allocated space becomes exhausted and it becomes necessary to delete blocks that are no longer needed. Utility routines are provided to list the identifiers of existing blocks and to delete blocks.

The `/LIST` command lists the identifiers of blocks existing in core. When called with no arguments, it lists all blocks—not usually a wise thing to do because of the large number of ISX procedures and other control blocks that are usually present. Called with two arguments, one of which is omitted or "...." meaning all, `/LIST` lists the identifiers of all matching blocks. For example, to list the identifiers of all blocks with blocktype DATA and all blocks with blockname REV, use the commands

```
> /list data
> /list .... rev
```

The `NAMETS` and `NAMEST` commands (Chapter 3) also list names of blocks (and can provide additional information), but only for time series and data matrix blocks, respectively.

`DELETE` removes time series (TS), data matrices (ST) and ISX procedures (PROCS) from core. To delete series <DATA REV> and <DATA HOUS>, matrices DEMO and DEM2, and procedure REVANL, type:

```
> delete ts data rev data hous
> delete st demo dem2
> delete procs revanl
```

EXECUTING SYSTEM COMMANDS

Many operating system commands can be executed from within ISX via the `/SYS` command. For example, if `MSG` sends a message to another user,

```
> /sys msg operator please detach my tape now
```

The return code from commands executed via `/SYS` is returned as ISX variable $XXX.

CHAPTER 9

Graphics

As is true elsewhere, a picture is worth a thousand (or more) words in statistical data analysis. So-called printer-plots, shown throughout this primer, can be produced quickly and inexpensively at the terminal and are adequate in many situations. But high-resolution graphics, produced on special graphical output devices, provides additional capabilities, such as:

- graphical output of quality adequate for management presentations and publication
- increased resolution
- different types of graphs, such as maps and pie charts

This chapter introduces STATLIB's high-resolution graphics package, GRPAK. After preliminaries, such as discussions of graphics devices and how to get started, we discuss how to direct STATLIB plots to a graphics device and how to modify and enhance them. Finally, we describe some special plotting commands.

PRELIMINARIES

STATLIB's graphical output may be produced on different types of devices, such as CRT terminals or flatbed plotters connected to your terminal, and COMs (computer output microfilmers), which generate microfilm, slides, hard-copy prints, etc. See the *GRPAK Manual* and local information for more detail on available devices. Note that GRPAK is available for use only under certain computer operating systems, including VM/CMS.

To specify the *graphical output device* you are using (or, in some cases, wish to have output directed to), use the GRDEV command. For example,

```
> grdev hp7221
```

defines the output device to be the Hewlett-Packard 7221 pen plotter.

To tell STATLIB to direct plotted output to GRPAK, you must issue the following command:

```
> option * graphics on
```

If you want to suppress the printer-plots, issue the command

```
> option * pplot off
```

To reset either of these, use the same command, replacing "on" or "off" by "*." Codes "graphics" and "pplot" may be abbreviated to two letters.

Once these preliminary commands have been issued, plotting can begin. The following commands produce a scatter plot of the series <DATA REV> versus <DATA MSG> on both the terminal and the COMp80 computer output microfilmer:

```
> grdev comp80
> option * gr on
> plotts time 1973 1 1975 12 data rev : data msg
```

The resulting graphical output is shown in Figure 9.1.

In the examples in the remainder of this chapter, it will be assumed that the commands GRDEV and OPTION * GR ON have been issued.

GRPAK continues to plot on the same "page" until GRBEG is called to begin a new one. GRFIN should be called to terminate graphics output before the end of the ISX session. A typical session, then, might include

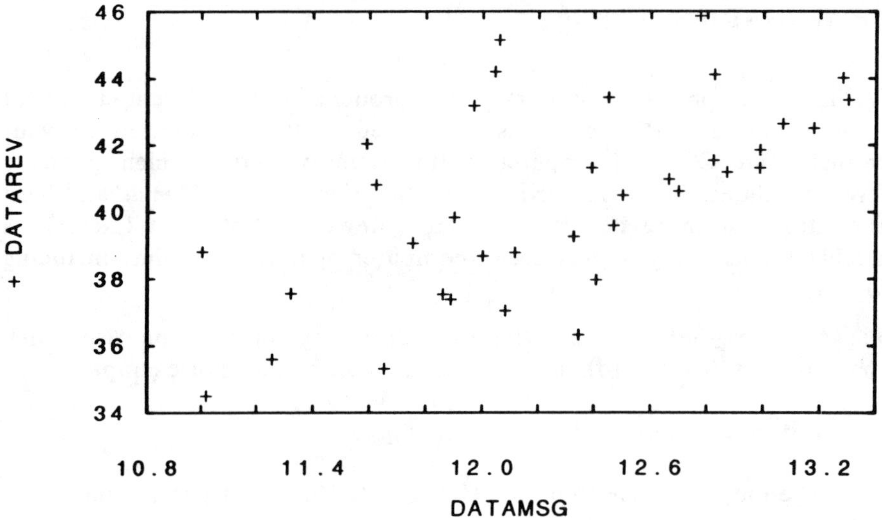

Figure 9.1. Scatter plot of <DATA REV> vs. <DATA MSG>

the following statements:

```
> grdev ..
> option * gr on
..
> plotts ..
> grbeg
..
> plotts ..
> grbeg
..
> plotst ..
> grfin
```

MODIFYING AND ENHANCING

Plots generated via the STATLIB plotting commands, as given in the previous section, may be modified by suppressing some of the features normally supplied, and enhanced by adding your own.

Suppressing Plot Portions

Each plot generated by the STATLIB plotting routines may be thought of as consisting of four parts:

- Scale. Scaling information (minima and maxima).
- Axes. Horizontal and vertical axes.
- Data. Points to be plotted.
- Labels. Names of variables or time series used to label the axes.

Each of these is normally on, that is, a call to a STATLIB plotting routine results in all of these items being included. You may control each, however, by calling `PARMGR` with codes GRSCALE, GRAXES, GRDATA, and GRLABELS, specifying ON, OFF, or ASIS. For example,

```
> parmgr grlabels off
> plotts time 1973 1 1975 12 data rev : data msg
```

produces the plot in Figure 9.2, identical to that in Figure 9.1 except that no axis labels are generated.

The ASIS option causes the corresponding portion of the plot to be drawn with minimal, if any, changes to graphics parameters; otherwise reasonable values are used. For example, specifying ASIS for GRLABELS causes the labels to be written using the current values (set via `PARMGR`) for character size and color, rather than selecting values.

`PARMGR` is GRPAK's parameter-setting routine. In addition to the parameters shown here, it can be used to set all the `PPLOTP` parameters (from Chapter 8), as well as many other GRPAK parameters. Any number of parameters can be set in a single call—For example:

```
> parmgr graxes off ; grlabels asis ; vmin 0
```

Scaling

Scaling is normally done automatically, that is, the horizontal and vertical minima and maxima are determined by the range of the data to be plotted. Scaling may be specified by calling `PARMGR` (or `PPLOTP`) to define scales for the STATLIB plotting routines; this determines the scaling for both printer-plots and graphical output:

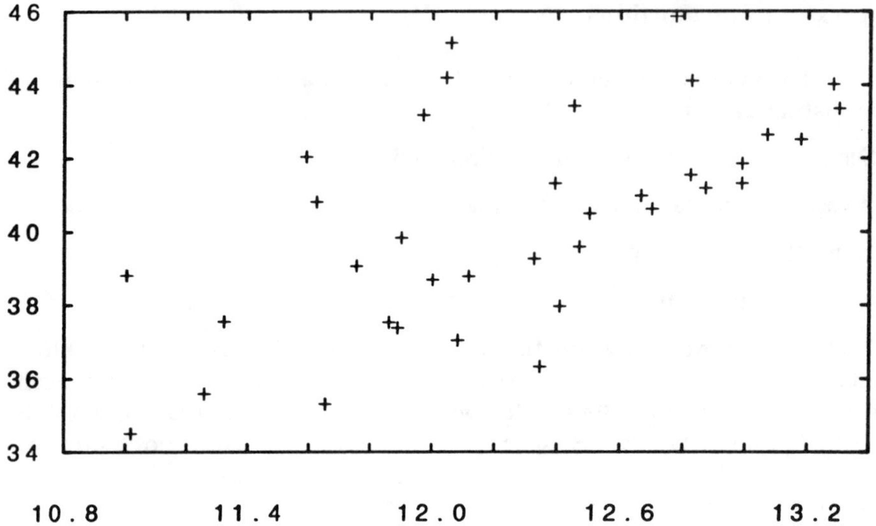

Figure 9.2. Scatter plot with labels suppressed.

```
> parmgr vmin 10 ; hmin 1973 ; hmax 1978
> plotts time 1973 3 1977 6 data rev
```

These commands produced the plot shown in Figure 9.3. **PARMGR** set the vertical minimum to 10 and the horizontal scale to 1973—1978. The series <DATA REV> was then plotted against time, with automatic scaling suppressed.

Drawing Axes

As with scaling, axes are usually drawn automatically. But, with **AXESGR**, you may draw axes in place of, or in addition to, those normally supplied.

In the following simple example, an additional vertical axis is drawn at the beginning of the year 1978 after plotting a time series:

```
> plotts time 1975 1 1978 12 data rev
> axesgr v loc 1978 nolabels
```

Specifying NOLABELS suppresses the tick-mark labels on this axis. To add a horizontal axis at rev = 55, with labels half their usual size, use:

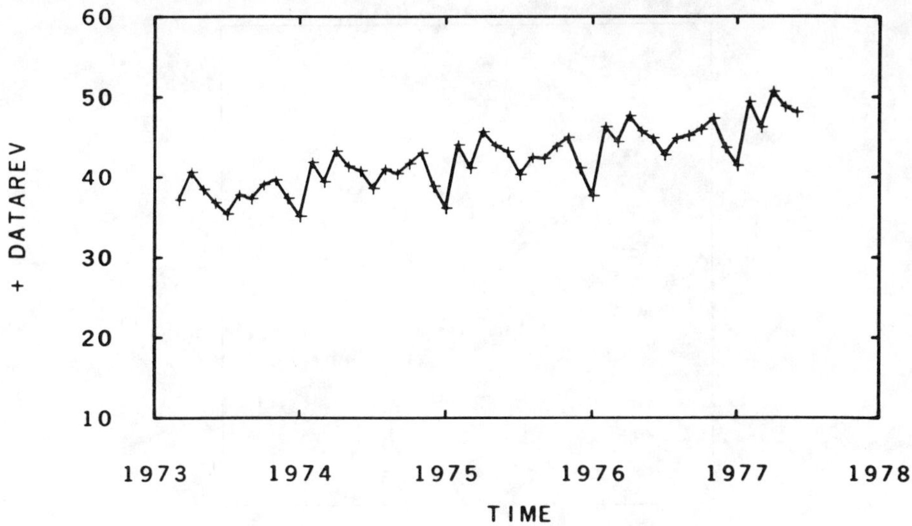

Figure 9.3. Time plot with user-defined scales.

```
> axesgr h loc 55 cex .5
```

Here "cex .5" sets the character expansion (of labels) to .5 for this axis.

Axes may be specified as *h* (horizontal) or *v* (vertical), as *x* or *y*, as *b*, *l*, *t*, or *r*, or by numbers representing sides (see Figure 9.4). Additional **AXESGR** parameters allow you to specify axis type (linear, time, log, etc.) and number of tick-mark intervals; if not specified, they default to the most recently used values.

In the following example, horizontal and vertical scaling and axes are specified by **PARMGR** and **AXESGR**; **PLOTTS** is called with axes suppressed:

```
> parmgr hmin 1974 ; hmax 1979 ; vmin 0 ; vmax 70
> parmgr graxes off
> plotts time 1975 1 1978 12 data rev
> axesgr h type time intervals 12 cex 1.5
> axesgr v type linear intervals 7 cex 1.3
```

The resulting plot is shown in Figure 9.5. The horizontal axis is a time axis with tick-marks every twelve points, since the series has twelve months per year, and tick-mark labels of size 1.5. The vertical axis is

Figure 9.4. Areas of a graph.

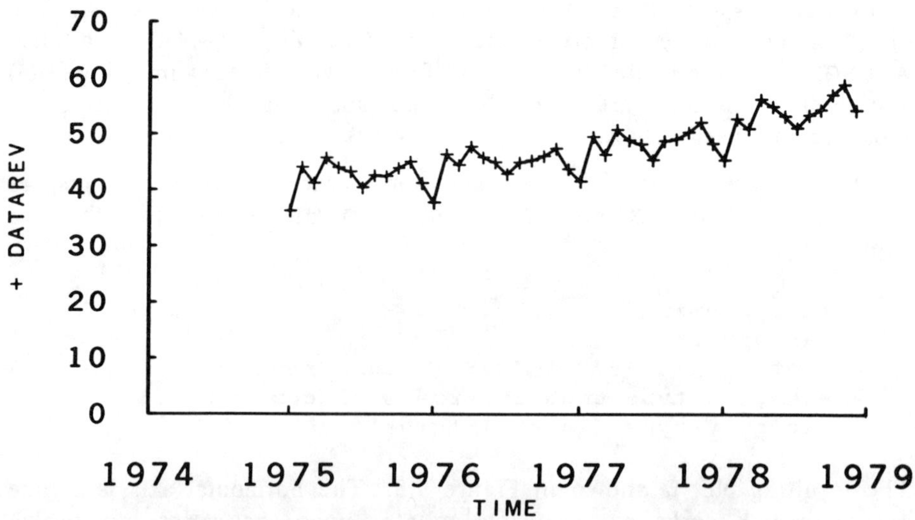

Figure 9.5. Time plot with pre-specified scaling and axes.

linear with seven intervals and labels of size 1.3.

Writing Labels and Text

Most STATLIB plotting routines do not provide an overall title. Also, axes are usually labeled in terms of data matrix variable names or time series identifiers. TEXTGR will replace these labels with others, or plot an arbitrary string of text at any location on the graph. The location can be specified by side and number of lines out from the plot boundary or by an *x-y* coordinate on the plot (see Figure 9.4).

The command

```
> textgr side b line 3 ; year
```

writes the word "YEAR" on the third line out from the bottom of the plot.

Additional TEXTGR parameters allow setting of such things as *character size* and *string rotation*.

In the following example, <DATA REV> is plotted against time, with automatic labeling suppressed and text added via TEXTGR. Output is shown in Figure 9.6.

```
> parmgr grlabels off
> plotts time 1974 1 1978 12 data rev
> textgr side 3 line 2 cex 1.5 ; revenue series
> textgr side 2 line 3 srt 90 ; dollars
> textgr xy 1976 55 cex .85 ; text at an x,y point
```

Here "REVENUE SERIES" is written in the top margin with size 1.5, "DOLLARS" is written along the *y*-axis with string rotation 90 (degrees), and "TEXT AT AN X,Y POINT" is written at the point (1976,55) with size .85.

Drawing Lines and Plotting Points

More detailed plot enhancements include drawing individual lines and points on a plot. As with GRPAK procedures already discussed, appropriate graphics parameters can be set for these calls.

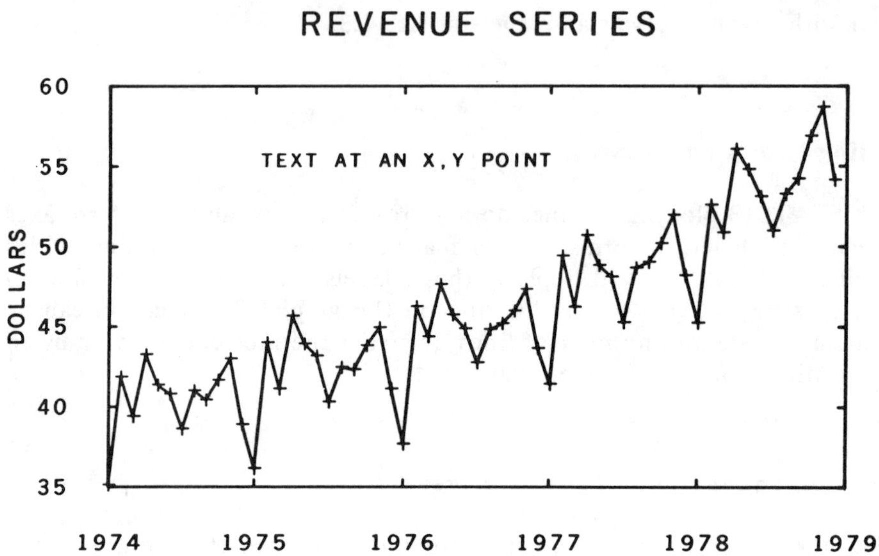

Figure 9.6. Time plot with added labels and text.

LINEGR draws a straight line:

```
> plotts time 1973 1 1977 12 data rev
> linegr 1973 43 1978 43 lty 2 col 3
```

In this example, a line is drawn from (1973,43) to (1978,43), after plotting <DATA REV> against time. Line type 2 and color 3 are used to draw the line.

POINTGR plots points:

```
> pointgr 1976 50 pch *
> pointgr 1976 37 pch :2 col 4
```

Here an asterisk (*) is plotted at (1976,50) and marker symbol 2 at (1976,37). *Marker symbols* are special plotting characters (bullets, triangles, etc.), specified by a colon followed by a code number.

The output from the above LINEGR and POINTGR calls is shown in Figure 9.7.

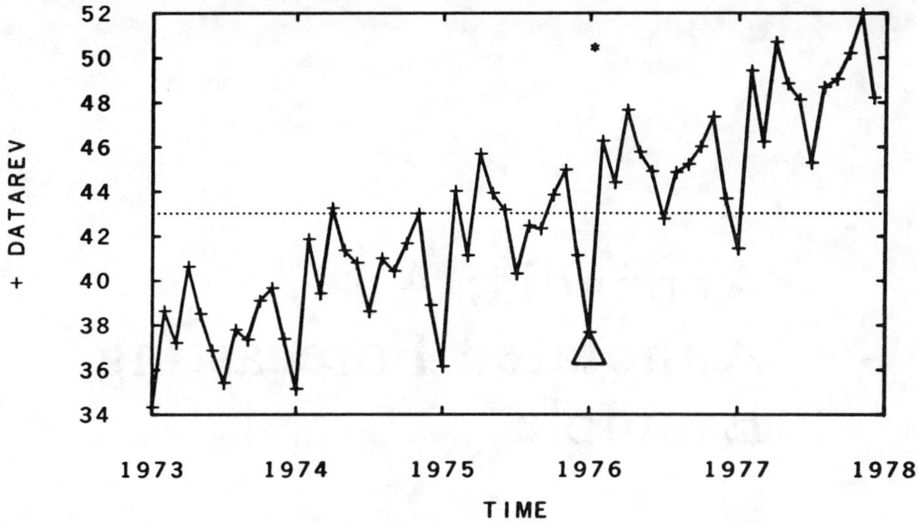

Figure 9.7. Straight line and points overlaid on time plot.

SUMMARY

As with other STATLIB packages, we have given here only a brief introduction to GRPAK. Graphics is by nature a very detailed and complex process; GRPAK has more commands, and GRPAK commands have many optional arguments that we have not tried to show or describe. GRPAK also has additional commands that generate specific plots, such as pie charts and bar charts.

Appendix: An Annotated Forecasting Example

Here is an annotated session that will guide you through a forecasting example using STATLIB. The session starts with inputting time series in the interactive ISX environment. In the next stage, several basic data manipulation and time series analysis techniques are introduced in preparation for modeling. These include transformations, differencing, autocorrelation analysis, and plotting. A simple linear regression model is built, using a robust option to protect against the presence of outliers. The model is further enhanced by incorporating a moving-average parameter in the sense of an ARIMA time series model. Finally, an *ex post* forecasting test is performed to assess the value of the model as a forecasting tool. These models are discussed in detail in Chapter 20 of *The Beginning Forecaster* and in Chapter 3 of *The Professional Forecaster*.

INPUTTING DATA

```
. isx
```

Command to enter ISX environment.

```
> term * dev prtr * pagsiz 60
```

Command to initialize system parameters for an interactive session, define the output device to be a printer (so plots will be enclosed in boxes), and set page width to sixty columns.

```
> input series .... ttmg .... hous
INPUT FROM SERIES DATA * ...
  DYNINP: 28 .DAT TTMG
  DYNINP: 28 DATA TTMG
  DYNINP: 28 .DAT HOUS
  DYNINP: 28 DATA HOUS
```

Command to load time series into ISX, followed by computer response indicating time control blocks (<.DAT TTMG>, <.DAT HOUS>) and data blocks (<DATA TTMG>, <DATA HOUS>) loaded.

```
> namets t data ttmg data hous
  DATA TTMG  JAN 1959-DEC 1978
  DATA HOUS  JAN 1959-DEC 1978
```

Command to identify time series and duration, followed by computer response.

DATA MANIPULATION AND ANALYSIS

```
> tsfnm data qtmg sum data ttmg 3 s
TSFNM:   DATA QTMG = SUM(DATA TTMG), JAN 1959-DEC 1978
         SEG LGTH = 3, NEW TIME = 1959:1-1978:4 <4 M/Y>
```

Command to transform monthly total telephone main gain series TTMG into quarterly main gain series QTMG by summing segments of three months. Original time series is <DATA TTMG>; quarterly series is stored as <DATA QTMG>.

```
> plotts time 1959 1 1978 4 data qtmg
TSPAK  / TSPLT

       |---|---|---|---|---|---|---|---|---|---
      |                                       |
      |                                       |
600000-                                      +-
      |                                     + |
      |                     +   +   +   +     |
      |                          ++  +   ++   |
400000-  ++                  ++       +       -
      |    +        +   + +                   |
      |+     +   +++  ++ +                    |
      | +     +            +       +          |
200000-        +               +       +      -
      |     +   +  +   +                      |
      |                                       |
      |                                    +  |
       |---|---|---|---|---|---|---|---|---|---
       1959    1961    1963    1965    1967

      1 + DATAQTMG  VS  TIME

TSPAK  / TSPLT

       |---|---|---|---|---|---|---|---|---|---
      |                                 +   + |
      |             +   +                 +   |
600000-  +        +  ++                  +   +-
      |      +                      +         |
      |+  +       +       + +   + +   +       |
      |    +   +         +       +   +        |
400000-       +  +                            -
      |                                       |
      |                       +            +  |
      | +              +     +                |
200000-     +      +       +           +      -
      |         +              +   +          |
      |                                       |
      |                                       |
       |---|---|---|---|---|---|---|---|---|---
       1969    1971    1973    1975    1977

      1 + DATAQTMG  VS  TIME
      1959:1-1978:4
```

Command to plot quarterly main gain series QTMG from first quarter 1959 through fourth quarter 1978. Series exhibits some trend, seasonality, and variability, increasing with time. Terminal response indicates that the series is plotted against time, using "+" as the plotting character.

```
> tran qtmg dqmg diff 4
TSDIFF: DATA DQMG = 4 DIFF OF DATA QTMG, 1960:1-1978:4
```

Command to calculate fourth-order differences of QTMG in order to remove seasonality and render series stationary. Differenced series is named <DATA DQMG>.

```
> plotts time 1960 1 1978 4 data dqmg
TSPAK  / TSPLT

        |---|---|---|---|---|---|---|---|---|-
       |                                      |
 200000-                                      -
       |                                     +|
       |        +        ++                   |
       |                   +               +  |
       |       +  +     +      ++       + + + |
       |+              +    + +  +     +      |
      0-     ++   + +              +   +      -
       |            +++      +     + +        |
       |                            +         |
       | ++                                   |
       |   +                             +    |
       |    +                                 |
-200000-                                      -
       |                                      |
        |---|---|---|---|---|---|---|---|---|-
        1960    1962    1964    1966    1968

       1 + DATADQMG  VS  TIME
```

```
TSPAK   / TSPLT

        --|---|---|---|---|---|---|---|---|---
       |            +                         |
 200000-                          +           -
       |          +              +       +    |
       |                                +  +  |
       |             +                    +   |
       |         + +                          |
       |               +        +   + ++      |
      0-+             + +          + +      ++-
       |    + ++           +   +              |
       | +++              +                   |
       |     +                                |
       |        +        +  +                 |
       |                                      |
-200000-                     +                -
       |                      +               |
        --|---|---|---|---|---|---|---|---|---
          1970    1972    1974    1976    1978

             1 + DATADQMG   VS   TIME
             1960:1-1978:4
```

Command to plot differenced series DQMG. Time plot shows stable variability and no trend, suggesting stationarity (mean and variance constant over time).

```
> tsfnm data qhos sum data hous 3 s
TSFNM:   DATA QHOS = SUM(DATA HOUS), JAN 1959-DEC 1978
         SEG LGTH = 3, NEW TIME = 1959:1-1978:4 <4 M/Y>
```

Command to transform monthly U.S. housing-starts (HOUS) into quarterly housing-starts QHOS by summing segments of three months. Original series is <DATA HOUS>, quarterly series <DATA QHOS>.

```
> tran qhos dqho diff 4
TSDIFF: DATA DQHO = 4 DIFF OF DATA QHOS, 1960:1-1978:4
```

Command to calculate fourth-order differences of quarterly housing-starts to remove trend and seasonal pattern, as was done with QTMG. Differenced series is named <DATA DQHO>.

```
> namets t data dqmg data dqho
  DATA DQMG   1960:1-1978:4    <4 M/Y>
  DATA DQHO   1960:1-1978:4    <4 M/Y>
```

Command to identify time series and duration, followed by computer response.

```
> tsxcor data dqho data dqmg mgho 1960 1 1978 4
TSXCOR: MGHO = DATA DQHO LEADING DATA DQMG, 25 LAGS
        1960:1-1978:4
> plotts time -24 1 24 1 xcor mgho
TSPAK  / TSPLT

   ----|---------|---------|---------|---------|----
 1-                        I                        -
  |                        I                        |
  |                        I                        |
  |                        I++                      |
  |                        +++++                    |
  |        +++            +I+++++             +     |
  |      ++++++           +I++++++           ++++   |
 0--+++++------+---------+--------+---------+----+++-
  |+            +++++++++  I       +++++++++        |
  |              +++++++   I        +++++++         |
  |                        I                        |
  |                        I                        |
  |                        I                        |
  |                        I                        |
-1-                        I                        -
   ----|---------|---------|---------|---------|----
      -20       -10        0         10        20

    1 + XCORMGHO   VS   TIME
    -24-24
```

Commands to calculate and plot the sample cross-correlation between DQMG and DQHO in order to determine a lead/lag relationship between the two. The "lead" relationship, indicated by the peak at 2 on the positive horizontal axis, suggests that housing-starts lead main gain by two quarters. Therefore, in the regression model the differenced housing-starts variable DQHO will be included as DQHO 2. Note that the cross-correlation function, <XCOR MGHO>, is plotted from minus twenty-four lags to plus twenty-four lags by specifying "time -24 1 24 1."

REGRESSION MODELING

```
> reg time 1960 1 1978 4 dqmg ; dqho 2
TSREG:   MODEL=L.S., TYPE=DATA, 1960:3-1978:4
         DEPV=DQMG

TSPRDN:  MODEL=L.S., TYPE=DATA/PRDN/RSID, 1960:3-1978:4
         DEPV=DQMG

TSUNDF:  PRDN QTMG = 4 UNDF OF PRDN DQMG, BASE = DATA QTMG
         1960:3-1978:4

TSFN2:   RSID QTMG = SUB(DATA QTMG,PRDN QTMG LAG 0)
         1960:3-1978:4

REG - REGR ANALYSIS..$RGA, RSID PLOT..$RP [IND [LAG]]
- RSID VS PRDN PLOT..$RVP, ACOR PLOT..$AP, PACF PLOT..$PAP
- PRTL RSID PLOT..$PRP IND [LAG]
- PRTL ADJ RSID PLOT..$PAR IND [LAG], REGR COEF CORR..$RCC
- RSID ANALYSIS..$RSA, CHI-SQUARED..$CSQ
- DURBIN-WATSON..$DW, RSID STD NORMAL PLOT..$NRM
```

Command to run a regression on the differenced main gain series as a function of differenced housing-starts lagged two quarters. Response indicates that the regression model was estimated for dependent variable DQMG and that various outputs can be obtained from the list of abbreviated optional-output commands.

```
> $rga
REGPAK / REGANL


   SAMPLE SIZE  . . . . . . . .    74
   SUM OF WEIGHTS . . . . . . .   7.4000D+01
   ESTIMATED STD DEV  . . . . .   8.0312D+04
   R SQUARED  . . . . . . . . .   0.2931


   VARIABLE      COEFFICIENT     ESTD STD DEV        T

    0              1.0172D+04      9.3485D+03      1.0880
    2 DQHO...2     5.8095D+02      1.0634D+02      5.4633

    1 DQMG....     DEPENDENT VARIABLE


   ANALYSIS OF VARIANCE

      SOURCE        DF        SS           MS            F

      REGRESSION     1     1.9252D+11   1.9252D+11     29.848
      ERROR         72     4.6441D+11   6.4501D+09

      TOTAL         73     6.5693D+11

TSREGA: MODEL=L.S., TYPE=DATA, DEPV=DQMG, 1960:3-1978:4
```

Command to obtain standard regression analysis output.

```
> $dw
DURBIN-WATSON STATISTIC (1ST ORDER) = 1.1578481
```

Command to calculate and display the Durbin-Watson statistic to test for first-order autocorrelation of the residuals. At the five-percent significance level and with seventy-three degrees of freedom, the lower and upper tabled Durbin-Watson values are 1.51 and 1.57, respectively. A comparison of the observed statistic, 1.16, with these tabled values suggests first-order positive autocorrelation of the residuals.

```
> $ap
TSACOR: RSID DQMG, 24 LAGS, 1960:3-1978:4

TSPAK  / TSPLT

   |---------|---------|---------|---------|---------|--------
 1-+                                                ‑
  |+                                                |
  |+                                                |
  |+                                                |
  |+ +                                              |
  |==========================+======================|
  |+ +           +         + + +                    |
 0-----+-------+---+-+-+-+---------------+-+-+-+-----
  |      + + +                   + + + +         + +|
  |========+=========================+==============|
  |                                                 |
  |                                                 |
  |                                                 |
  |                                                 |
-1-                                                 -
   |---------|---------|---------|---------|---------|--------
   0         5         10        15        20

  1 + ACORDQMG  VS  TIME
  0-24
```

Command to generate an autocorrelation plot of the residuals. Notice positive first-order autocorrelation which substantiates the Durbin-Watson test results.

```
> $rp
TSPAK  / TSPLT
```

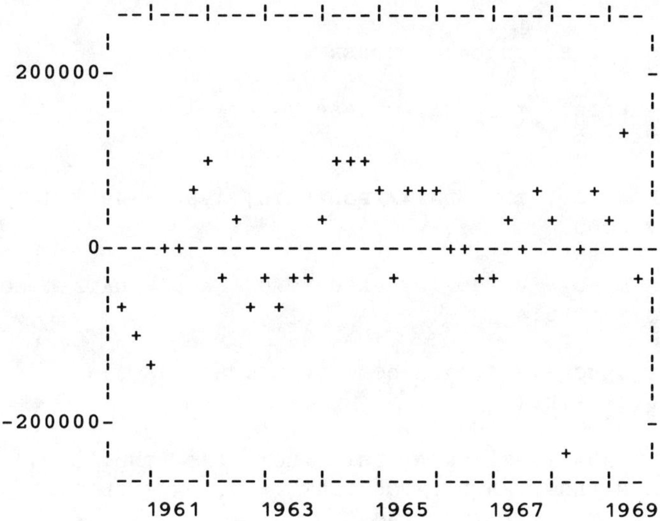


```
1 + RSIDDQMG  VS  TIME
```

```
TSPAK  / TSPLT
```

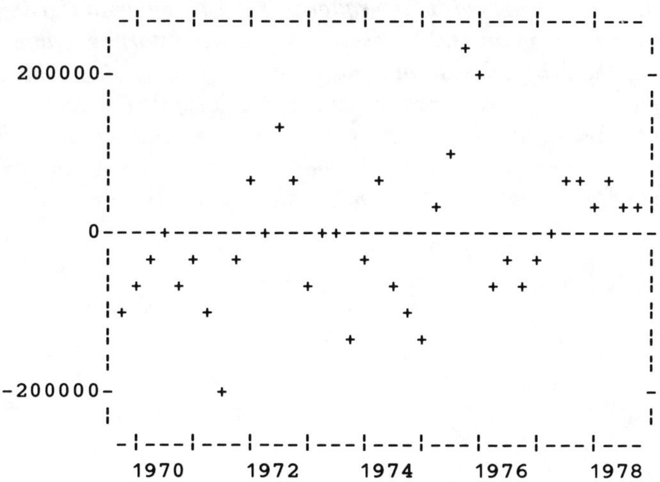


```
1 + RSIDDQMG  VS  TIME
1960:3-1978:4
```

Command to generate a time plot of the residuals to check for constant variance, randomness, and outliers. This is one of several diagnostic tools available for checking the validity of regression assumptions.

```
> rwrreg dqmg ; dqho 2
6 ITERATIONS ( 2 BISQUARE, 4 HUBER )

TSREG:   MODEL=L.S., TYPE=DATA, 1960:3-1978:4
         DEPV=DQMG

TSPRDN: MODEL=L.S., TYPE=DATA/PRDN/RSID, 1960:3-1978:4
         DEPV=DQMG

TSUNDF: PRDN QTMG = 4 UNDF OF PRDN DQMG, BASE = DATA QTMG
         1960:3-1978:4

TSFN2:   RSID QTMG = SUB(DATA QTMG,PRDN QTMG LAG 0)
         1960:3-1978:4

RWRREG - REGR ANALYSIS..$RGA, RSID PLOT..$RP [IND [LAG]]
- RSID VS PRDN PLOT..$RVP, ACOR PLOT..$AP, PACF PLOT..$PAP
- PRTL RSID PLOT..$PRP IND [LAG]
- PRTL ADJ RSID PLOT..$PAR IND [LAG], REGR COEF CORR..$RCC
- RSID ANALYSIS..$RSA, CHI-SQUARED..$CSQ, DURBIN-WATSON..$DW
- RSID STD NORMAL PLOT..$NRM
```

Command to perform a weighted least squares robust regression using M-estimators. Beginning with an OLS model to obtain starting values for coefficient estimates, the procedure iterates, using a Huber weighting scheme for four iterations followed by two iterations using the bisquare scheme. This procedure removes the harmful effects of outliers on OLS estimation by assigning weights to each data value. Probable outliers are given smaller weights than other values, and thus have less influence on the final regression estimates.

```
> $rga
REGPAK / REGANL


  SAMPLE SIZE  . . . . . . . . .   74
  SUM OF WEIGHTS . . . . . . . .  6.9925D+01
  ESTIMATED STD DEV  . . . . .   7.3410D+04
  R SQUARED  . . . . . . . . . .  0.3592


  VARIABLE       COEFFICIENT      ESTD STD DEV          T

   0              1.0258D+04       8.5446D+03        1.2006
   2 DQHO...2     6.2407D+02       9.8244D+01        6.3523

   1 DQMG....     DEPENDENT VARIABLE


  ANALYSIS OF VARIANCE

      SOURCE          DF         SS              MS             F

      REGRESSION       1     2.1746D+11     2.1746D+11     40.352
      ERROR           72     3.8801D+11     5.3890D+09

      TOTAL           73     6.0547D+11

 TSREGA: MODEL=L.S., TYPE=DATA, DEPV=DQMG, 1960:3-1978:4
```

Command to obtain standard regression analysis output from the weighted least squares robust regression. The SUM OF WEIGHTS (69.93) is less than the SAMPLE SIZE (74), since some data values are downweighted by the procedure. Notice that the coefficient for DQHO from the robust regression is within one standard deviation of the estimated OLS coefficient. This suggests that any existing outliers are not having a major impact on the estimated coefficients. Note: Summary statistics, such as R-squared *and* t *statistics, are generally not valid for use in significance testing in this case.*

TIME SERIES FORECASTING

```
> acplt both data qtmg
TSACOR: DATA QTMG, 26 LAGS, 1959:1-1978:4

TSPAK  / TSPLT

      ¦---------¦---------¦------
    1-+                          -
     ¦+                          ¦
     ¦+   +                      ¦
     ¦+   +   +   +   +          ¦
     ¦+   +   +   +   +   +   +  ¦
     ¦===========================¦
     ¦+ + +   +   +   +   +   +  ¦
    0--+-+--+---+---+---+---+-----
     ¦     + + + + + + + + + + ++¦
     ¦===========================¦
     ¦                           ¦
     ¦                           ¦
     ¦                           ¦
     ¦                           ¦
   -1-                           -
      ¦---------¦---------¦------
      0        10        20

     1 + ACORQTMG  VS  TIME
     0-26
```

```
TSPACF: DATA QTMG, 26 LAGS, 1959:1-1978:4

TSPAK  / TSPLT

      |---------|---------|------
    1-+                          -
      |+                         |
      |+  +                      |
      |+  +                      |
      |+  +                      |
      |==========================|
      |+ + +   +   ++  +  +      |
    0--+-+---+-+-+--++--+--++++++-
      |    ++   +      +  +      |
      |==========================|
      |    +                     |
      |                          |
      |                          |
      |                          |
   -1-                           -
      |---------|---------|------
      0        10        20

       1 + PACFQTMG  VS  TIME
       0-26
```

Command to generate both a sample autocorrelation (ACOR) plot and a sample partial autocorrelation (PACF) plot for quarterly main gain series QTMG. Terminal response indicates blocktype and blockname of series, number of lags, and time frame. Autocorrelation plot exhibits seasonality, suggested by slowly decaying spikes (characteristic of a nonstationary time series) at four-lag intervals. Terminal response indicates that autocorrelations of QTMG are plotted for lags 0 through 26. Plot of partial autocorrelations confirms strong seasonality by spike at lag four.

```
> acplt both data dqmg
TSACOR: DATA DQMG, 25 LAGS, 1960:1-1978:4

TSPAK  / TSPLT

     |---------|---------|-----
   1-+                         -
    |+                         |
    |+                         |
    |++                        |
    |++                        |
    |==+=======================|
    |++++        +  +          |
   0--------------++-++++----+-
    |    +++++++++       ++++  |
    |==========================|
    |                          |
    |                          |
    |                          |
    |                          |
  -1-                          -
     |---------|---------|-----
     0         10        20

   1 + ACORDQMG  VS  TIME
   0-25
```

```
TSPACF: DATA DQMG, 25 LAGS,  1960:1-1978:4

TSPAK  / TSPLT

   ¦---------¦---------¦-----
 1-+                         -
  ¦+                         ¦
  ¦+                         ¦
  ¦++                        ¦
  ¦++                        ¦
  ¦========================+=¦
  ¦++   +       +  +       + ¦
 0---++--++-++-+----++-+++----
  ¦    +   +  +  ++   +   + +¦
  ¦========+=================¦
  ¦    +                     ¦
  ¦                          ¦
  ¦                          ¦
  ¦                          ¦
-1-                          -
   ¦---------¦---------¦-----
   0         10        20

  1 + PACFDQMG   VS   TIME
  0-25
```

Command to generate autocorrelation (ACOR) and partial autocorrelation (PACF) plots of differenced main gain series. The autocorrelation plot exhibits sinusoidal decay, which is characteristic of a low-order autoregressive model. The partial autocorrelation plot supports the preliminary identification of autoregressive structure—The spike at lag one suggests a first-order autoregressive model for the differenced series.

```
> regpar con off
```

Command to suppress constant term in regression model. This is done because the series is assumed to be stationary. If the mean of the residuals in a tentative model is significantly different from zero, a "deterministic trend constant" should be included. Since this is generally the last parameter to be included in a model-building cycle, the constant term is suppressed at this stage.

```
> reg dqmg 1
TSREG:   MODEL=L.S., TYPE=DATA, 1960:2-1978:4
         DEPV=DQMG

TSPRDN:  MODEL=L.S., TYPE=DATA/PRDN/RSID, 1960:2-1978:4
         DEPV=DQMG

TSUNDF:  PRDN QTMG = 4 UNDF OF PRDN DQMG, BASE = DATA QTMG
         1960:2-1978:4

TSFN2:   RSID QTMG = SUB(DATA QTMG,PRDN QTMG LAG 0)
         1960:2-1978:4

REG - REGR ANALYSIS..$RGA, RSID PLOT..$RP [IND [LAG]]
- RSID VS PRDN PLOT..$RVP, ACOR PLOT..$AP, PACF PLOT..$PAP
- PRTL RSID PLOT..$PRP IND [LAG]
- PRTL ADJ RSID PLOT..$PAR IND [LAG], REGR COEF CORR..$RCC
- RSID ANALYSIS..$RSA, CHI-SQUARED..$CSQ
- DURBIN-WATSON..$DW, RSID STD NORMAL PLOT..$NRM
```

Command to run a regression of the fourth-differenced main gain series against itself lagged one quarter (first-order autoregressive model). The REG *procedure automatically undifferences series DQMG to create predictions in the original domain of series QTMG. For diagnostic purposes, residuals are created automatically and are named <RSID QTMG>.*

```
> $rga
REGPAK / REGANL


  SAMPLE SIZE  . . . . . . . .    75
  SUM OF WEIGHTS . . . . . . .   7.5000D+01
  ESTIMATED STD DEV  . . . . .   7.8094D+04
  R SQUARED  . . . . . . . . .   0.3341


  VARIABLE       COEFFICIENT      ESTD STD DEV          T

   2 DQMG...1     5.7752D-01       9.4778D-02       6.0934

   1 DQMG....    DEPENDENT VARIABLE


  ANALYSIS OF VARIANCE

     SOURCE         DF        SS              MS             F

     REGRESSION      1    2.2644D+11     2.2644D+11      37.129
     ERROR          74    4.5130D+11     6.0986D+09

     TOTAL          75    6.7774D+11

TSREGA: MODEL=L.S., TYPE=DATA, DEPV=DQMG, 1960:2-1978:4
```

Command to obtain standard regression analysis output for first-order autoregressive model of the fourth-differenced main gain series DQMG.

```
> $csq
TSACOR: RSID DQMG, 25 LAGS, 1960:2-1978:4
CHI-SQUARE = 30.1383694 ( 24 DF )
```

Command to calculate and display the chi-square statistic for randomness of the residuals. At the five-percent significance level and with twenty-four degrees of freedom, the tabled value is 36.42. A comparison with the observed statistic of 30.1 indicates that we cannot reject the hypothesis that the residuals are random. However, further structure may still be uncovered by analyzing the residual autocorrelation and partial autocorrelation functions.

```
> $ap
TSACOR: RSID DQMG, 25 LAGS, 1960:2-1978:4

TSPAK  / TSPLT

        |---------|---------|-----
     1-+                          -
       |+                         |
       |+                         |
       |+                         |
       |+                         |
       |==========================|
       |+  +         +  +  +      |
     0--++---++--+---+--++-+-+-++-
       |    ++  ++ ++  +     + +  |
       |====+=====================|
       |                          |
       |                          |
       |                          |
       |                          |
    -1-                           -
        |---------|---------|-----
        0        10        20

       1 + ACORDQMG   VS   TIME
       0-25
```

Command to generate a residual autocorrelation plot. The plot exhibits a barely significant spike at lag four followed by truncation. This suggests the need for a moving-average term of order four in the model. There is a one in twenty chance that a spike will lie outside the dashed limits.

```
> $pap
TSPACF: RSID DQMG, 25 LAGS, 1960:2-1978:4

TSPAK  / TSPLT

   ¦---------¦---------¦-----
 1-+                         -
  ¦+                         ¦
  ¦+                         ¦
  ¦+                         ¦
  ¦+                         ¦
  ¦==========================¦
  ¦+  +   +     ++         + ¦
 0--++--++---+-----+++-+++----
  ¦    +   ++ ++  +   +   + +¦
  ¦========+======+=======+==¦
  ¦    +                     ¦
  ¦                          ¦
  ¦                          ¦
  ¦                          ¦
-1-                          -
   ¦---------¦---------¦-----
   0         10        20

  1 + PACFDQMG  VS  TIME
  0-25
```

Command to generate a residual partial autocorrelation plot. The spike at lag four, followed by decay at lags eight, twelve, and sixteen, confirms the need for a fourth-order moving-average term in the model.

```
> reg dqmg 1 ; m.a. 4
TSREG:   MODEL=L.S., TYPE=DATA, 1961:1-1978:4
         DEPV=DQMG

TSPRDN:  MODEL=L.S., TYPE=DATA/PRDN/RSID, 1961:1-1978:4
         DEPV=DQMG

TSUNDF:  PRDN QTMG = 4 UNDF OF PRDN DQMG, BASE = DATA QTMG
         1962:1-1978:4

TSFN2:   RSID QTMG = SUB(DATA QTMG,PRDN QTMG LAG 0)
         1962:1-1978:4

REG - REGR ANALYSIS..$RGA, RSID PLOT..$RP [IND [LAG]]
- RSID VS PRDN PLOT..$RVP, ACOR PLOT..$AP, PACF PLOT..$PAP
- PRTL RSID PLOT..$PRP IND [LAG]
- PRTL ADJ RSID PLOT..$PAR IND [LAG], REGR COEF CORR..$RCC
- RSID ANALYSIS..$RSA, CHI-SQUARED..$CSQ
- DURBIN-WATSON..$DW, RSID STD NORMAL PLOT..$NRM
```

Command to rerun the autoregressive model, including the moving-average term of order four.

```
> $rga
REGPAK / REGANL


   SAMPLE SIZE . . . . . . . . .   68
   SUM OF WEIGHTS . . . . . . . . 6.8000D+01
   ESTIMATED STD DEV . . . . .    7.1472D+04
   R SQUARED . . . . . . . . .    0.4420


   VARIABLE        COEFFICIENT      ESTD STD DEV          T

    2 DQMG...1      6.4327D-01       9.9755D-02       6.4485
    3 M.A....4     -4.9301D-01       1.1219D-01      -4.3943

    1 DQMG....     DEPENDENT VARIABLE


   ANALYSIS OF VARIANCE

       SOURCE        DF          SS              MS            F

       REGRESSION     2     2.6706D+11     1.3353D+11     26.140
       ERROR         66     3.3715D+11     5.1083D+09

       TOTAL         68     6.0421D+11

TSREGA: MODEL=L.S., TYPE=DDTA, DEPV=DQMG, 1962:1-1978:4
```

Command to obtain standard regression analysis output for the autoregressive model with a moving-average term.

```
> $csq
TSACOR: RSID DQMG, 22 LAGS, 1962:1-1978:4
CHI-SQUARE = 28.5805534 ( 20 DF )
```

Command to calculate and display the chi-square statistic. At the five-percent significance level and with twenty degrees of freedom, the tabled value is 31.4. Comparison with the observed statistic, 28.58, suggests that the residuals might be random.

```
> $rp
TSPAK   / TSPLT

        ¦---¦---¦---¦---¦---¦---¦---¦---¦-
       ¦                          +       ¦
       ¦                                  ¦
 100000-+             +            +      -
       ¦        + +                       ¦
       ¦  +      +     +  +    +          ¦
       ¦    + +     +   +     + +    +    ¦
      0--------+---+---------+------+-+--+-
       ¦                 + ++          ++ ¦
       ¦   + +       +                    ¦
       ¦ +                                ¦
-100000-                                  -
       ¦                                  ¦
       ¦                                  ¦
       ¦                                  ¦
-200000-                         +        -
        ¦---¦---¦---¦---¦---¦---¦---¦---¦-
        1962    1964    1966    1968    1970

       1 + RSIDDQMG  VS  TIME

TSPAK   / TSPLT

        --¦---¦---¦---¦---¦---¦---¦---¦---
       ¦                                  ¦
       ¦     ++ +          +        +     ¦
 100000-                             +    -
       ¦                                  ¦
       ¦            +        ++ + +    +  ¦
       ¦  +      +               +    + +¦
      0-----------++--++----+-----------+--
       ¦+  +                       +      ¦
       ¦       +                          ¦
       ¦                                  ¦
-100000- +              +      +          -
       ¦    +        +    +               ¦
       ¦                                  ¦
       ¦                 +                ¦
-200000-                                  -
        --¦---¦---¦---¦---¦---¦---¦---¦---
          1971    1973    1975    1977

       1 + RSIDDQMG  VS  TIME
       1962:1-1978:4
```

Command to generate a time plot of the residuals. The residual plot appears random with the possible exception of two outlying values in 1968.

We now rerun the final model in order to perform ex post *forecast tests, using the* FUPD *command.*

```
> fupd l.s. dqmg 1977 1978 1 68 8
FUPD: MODEL L.S. .. 1977 - 1978 BY 1 .. FIT 68 .. FCST 8

TSREG:   MODEL=L.S., TYPE=DATA, 1961:1-1976:4
         DEPV=DQMG

TSPRDN:  MODEL=L.S., TYPE=DATA/PRDN/RSID, 1961:1-1976:4
         DEPV=DQMG

TSUNDF:  PRDN QTMG = 4 UNDF OF PRDN DQMG, BASE = DATA QTMG
         1962:1-1976:4

TSFN2:   RSID QTMG = SUB(DATA QTMG,PRDN QTMG LAG 0)
         1962:1-1976:4

TSPRDN:  MODEL=L.S., TYPE=DATA/PRDN/RSID, 1977:1-1978:4
         DEPV=DQMG

TSUNDF:  PRDN QTMG = 4 UNDF OF PRDN DQMG, BASE = DATA QTMG
         1962:1-1978:4

TSFN2:   RSID QTMG = SUB(DATA QTMG,PRDN QTMG LAG 0)
         1962:1-1978:4

TSUNDF:  PRDN QTMG = 4 UNDF OF PRDN DQMG, BASE = DATA QTMG
         1962:1-1978:4

TSFN2:   RSID QTMG = SUB(DATA QTMG,PRDN QTMG LAG 0)
         1962:1-1978:4
```

```
TSPAK / TSRSDA

QTMG        DATA       PRDN        RESIDUAL      %      SQRT(ASR)

1977:1    520828.    474835.07    45992.93     8.83    45992.93
1977:2    193675.    180262.61    13412.39     6.93    33876.55
1977:3    686848.    551598.51   135249.49    19.69    82840.54
1977:4    607449.    427212.54   180236.46    29.67   115187.72

         2008800.   1633908.74   374891.26    18.66   115187.72


1978:1    628897.    467602.98   161294.02    25.65   125768.52
1978:2    314386.    175623.52   138762.48    44.14   128025.80
1978:3    698346.    548622.73   149723.27    21.44   131345.07
1978:4    596196.    425303.70   170892.30    28.66   136914.60

         2237825.   1617152.92   620672.08    27.74   155637.42


                                              23.44   136914.60

TSREG:   MODEL=L.S., TYPE=DATA, 1962:1-1977:4
         DEPV=DQMG

TSPRDN:  MODEL=L.S., TYPE=DATA/PRDN/RSID, 1962:1-1977:4
         DEPV=DQMG

TSUNDF:  PRDN QTMG = 4 UNDF OF PRDN DQMG, BASE = DATA QTMG
         1963:1-1977:4

TSFN2:   RSID QTMG = SUB(DATA QTMG,PRDN QTMG LAG 0)
         1963:1-1977:4

REG - REGR ANALYSIS..$RGA, RSID PLOT..$RP [IND [LAG]]
- RSID VS PRDN PLOT..$RVP, ACOR PLOT..$AP, PACF PLOT..$PAP
- PRTL RSID PLOT..$PRP IND [LAG]
- PRTL ADJ RSID PLOT..$PAR IND [LAG], REGR COEF CORR..$RCC
- RSID ANALYSIS..$RSA, CHI-SQUARED..$CSQ
- DURBIN-WATSON..$DW, RSID STD NORMAL PLOT..$NRM

TSPRDN:  MODEL=L.S., TYPE=DATA/PRDN/RSID, 1978:1-1979:4
         DEPV=DQMG

TSUNDF:  PRDN QTMG = 4 UNDF OF PRDN DQMG, BASE = DATA QTMG
         1963:1-1979:4
```

```
TSFN2:  RSID QTMG = SUB(DATA QTMG,PRDN QTMG LAG 0)
        1963:1-1978:4

FCST - RSID ANALYSIS..$RSA, PRINT FCSTS..$FPR ( FOR QTMG )

TSUNDF: PRDN QTMG = 4 UNDF OF PRDN DQMG, BASE = DATA QTMG
        1963:1-1979:4

TSFN2:  RSID QTMG = SUB(DATA QTMG,PRDN QTMG LAG 0)
        1963:1-1978:4

TSPAK / TSRSDA

QTMG         DATA         PRDN       RESIDUAL       %      SQRT(ASR)

1978:1     628897.     605577.63     23319.37      3.71     23319.37
1978:2     314386.     261823.12     52562.88     16.72     40661.10
1978:3     698346.     653974.83     44371.17      6.35     41934.27
1978:4     596196.     533052.17     63143.83     10.59     48121.18

          2237825.   2054427.76     183397.24      8.20     48121.18


                                                   8.20     48121.18

FUPD - REGR ANALYSIS..$RGA
```

Command to generate ex post forecast tests. L.S. is the model name by default. DQMG is the dependent variable. 1977 is year to begin forecast testing, 1978 the last year for forecast testing, and 1 the increment (in years). 68 is the (maximum) number of data values to use for model estimation and 8 is the number of periods (quarters) to forecast at each step.

At each step, the output from subroutine TSRSDA *provides a residual analysis, that is, a summary including forecast values (PRDN), forecast errors (RESIDUAL), and percent misses (%). Notice that the forecast values have been "undifferenced" automatically and are predictions of the raw series QTMG, not the differenced series DQMG.*

Results are shown for one and two years ahead, in the first case for 1977 and 1978. A summary showing overall percent miss is given for each year separately, and for both years together, to give an indication of the expected error one and two years out.

The process is repeated by updating the model with another year's data. Forecasts are derived by reestimating the model parameters, using data through 1977, and calculating predictions for 1978 and 1979.

The one- and two-line messages generated in the course of this analysis could be suppressed using OPTION, *as discussed in Chapter 8.*

```
> oo off
```

Command to suppress system procedure optional-output lists.

```
> reg dqmg 1 ; m.a. 4
TSREG:   MODEL=L.S., TYPE=DATA, 1961:1-1978:4
         DEPV=DQMG

TSPRDN:  MODEL=L.S., TYPE=DATA/PRDN/RSID, 1961:1-1978:4
         DEPV=DQMG

TSUNDF:  PRDN QTMG = 4 UNDF OF PRDN DQMG, BASE = DATA QTMG
         1962:1-1978:4

TSFN2:   RSID QTMG = SUB(DATA QTMG,PRDN QTMG LAG 0)
         1962:1-1978:4

> fcst thru 1980 4
TSPRDN:  MODEL=L.S., TYPE=DATA/PRDN/RSID, 1979:1-1980:4
         DEPV=DQMG

TSUNDF:  PRDN QTMG = 4 UNDF OF PRDN DQMG, BASE = DATA QTMG
         1962:1-1980:4

TSFN2:   RSID QTMG = SUB(DATA QTMG,PRDN QTMG LAG 0)
         1962:1-1978:4

> clim prdn arfs
TSARFS:  MODEL=L.S., VBL=DQMG, 1979:1-1980:4

TSCLIM:  DQMG, UCL/LCL = PRDN +/- 2 * ARFS
```

Commands to produce model predictions and confidence limits on future values. After REG *reruns the regression,* FCST *computes model predictions and* CLIM *computes 95-percent confidence limits through the end of 1980. Notice that the routine automatically undifferences the dependent variable DQMG to obtain forecasts in the original domain.*

```
> pplotp char 4 X+--
> pplotp vaxis 1978.875 :
> plotts time 1976 1 1980 4 data dqmg & prdn dqmg $
>   & lcl dqmg & ucl dqmg
TSPAK  / TSPLT

       |-------|-------|-------|-------|------
       |X                     :               |
200000-                       :        - - - --
       |+            X        :    - -        |
       |           X          :- -            |
       |               X X    :               |
       |  +      +   + + +    :               |
       |         X            :               |
       |    X   X   +      +  :               |
     0-   X + X +          X X:+   + + + + + +-
       |                     +: +             |
       |                      :               |
       |                      :               |
       |                      :               |
       |                      :-              |
       |                      : - - - - - - - |
       |-------|-------|-------|-------|------
       1976    1977    1978    1979    1980

      1 X DATADQMG  2 + PRDNDQMG  3 - LCL DQMG
      4 - UCL DQMG  VS  TIME
      1976:1-1980:4
```

Commands to plot predictions (fitted values before 1979), confidence limits, and some historical values for DQMG. PPLOTP *changes the first four plotting characters to "X" for actuals, "+" for predictions, "-" for lower confidence limits, and "-" for upper confidence limits; then draws a vertical axis, using plotting character ":" to separate past from future.*

Bibliography

[numbers in brackets refer to page numbers in this book]

AGARWALA, P., W. M. BRELSFORD, J. INGLIS, and D. B. PRESTON (1982). *STATLIB/ISX System Procedures.* To be published. [iv, 37, 38, 50, 76, 93, 128, 176]

This manual documents the System Procedures, an extension of the basic STATLIB system.

BRELSFORD, W. M., and D. A. RELLES (1981). *STATLIB—A Statistical Computing Library.* Englewood Cliffs, NJ: Prentice-Hall. [iv, 2, 32, 39, 133, 134, 176]

This manual contains documentation for the basic STATLIB system. It is organized in four parts:

Part I is an introduction to the library. It describes the various programming levels available, and presents, through examples, some of the library's principal packages.

Part II describes the library's high-level routines, all available in the STATLIB command language.

Part III describes the library's low-level FORTRAN routines, used as building blocks for the command language.

Part IV covers the various implementations of STATLIB and shows the operating system commands required to invoke parts of it.

CLEARY, J. P., AND H. LEVENBACH (1982). *The Professional Forecaster.* Belmont, CA: Lifetime Learning Publications. [v, 146]

LEVENBACH, H, AND J. P. CLEARY (1981). *The Beginning Forecaster.* Belmont, CA: Lifetime Learning Publications. [v, 117, 146]

These companion volumes use STATLIB extensively for examples, illustrations, models, and analyses. They are designed for the practicing forecaster. Both volumes stress a process for effective forecasting, with an emphasis on analyzing data before starting model-building. Because of its flexibility and power, this step is readily accomplished with STATLIB.

STATLIB/ISX GRAPHICS PACKAGE (GRPAK) (1982). Unpublished Bell System documentation. [137, 176]

Index

[command documentation references are included in brackets: *GM* (*GRPAK Manual*), *PM* (*Procedure Manual*), or *SM* (*STATLIB Manual*) followed by chapter and section numbers]
